Manufacturing System Throughput Excellence

Analysis, Improvement, and Design

Manufacturing System Throughput Excellence

Analysis, Improvement, and Design

Herman Tang
Eastern Michigan University
Michigan, USA

For general information on our other products and services or for technical support, please contact our Customer Care Department within the United States at (800) 762-2974, outside the United States at (317) 572-3993 or fax (317) 572-4002.

Wiley also publishes its books in a variety of electronic formats. Some content that appears in print may not be available in electronic formats. For more information about Wiley products, visit our website at www.wiley.com.

Library of Congress Cataloging-in-Publication Data

Names: Tang, He (Herman), author. | John Wiley & Sons, publisher.
Title: Manufacturing system throughput excellence : analysis, improvement, and design / Herman Tang.
Description: Hoboken, New Jersey : Wiley, [2024] | Includes index.
Identifiers: LCCN 2024008442 (print) | LCCN 2024008443 (ebook) | ISBN 9781394190324 (hardback) | ISBN 9781394190331 (adobe pdf) | ISBN 9781394190348 (epub)
Subjects: LCSH: Process control. | Production management. | Production engineering. | Industrial engineering.
Classification: LCC TS156.8 .T365 2024 (print) | LCC TS156.8 (ebook) | DDC 658.5–dc23/eng/20240402
LC record available at https://lccn.loc.gov/2024008442
LC ebook record available at https://lccn.loc.gov/2024008443

Cover Design: Wiley
Cover Image: © Edwin Tan/Getty Images

Set in 9.5/12.5pt STIXTwoText by Straive, Chennai, India

Contents

List of Figures

List of Tables

About the Author

Dr. Herman Tang is a full professor in the School of Engineering at Eastern Michigan University. With over 16 years of hands-on industry experience at Chrysler (now Stellantis N.V.), he worked extensively on vehicle manufacturing system development, product launches at plants, and production improvement. Dr. Tang holds a PhD in Mechanical Engineering from the University of Michigan – Ann Arbor, an MBA in Industrial Management from Baker College, and Mechanical Engineering degrees from Tianjin University. He is the author of four books: *Automotive Vehicle Assembly Processes and Operations Management* (SAE, 2017), *Manufacturing System and Process Development for Vehicle Assembly* (SAE, 2018), *Engineering Research – Design, Methods, and Publication* (Wiley, 2021), and *Quality Planning and Assurance – Principles, Approaches, and Methods for Product and Service Development* (Wiley, 2022). In 2022, he received the Eastern Michigan University's highest faculty honor, the Ronald W. Collins Distinguished Faculty Award.

Preface

The primary objective and responsibility for manufacturing managers and engineers are to achieve throughput excellence. This involves establishing clear and measurable objectives and applying scientific principles and methods to attain them. Throughput excellence is not a onetime achievement but a continuous pursuit that demands commitment, guidance, and constant effort.

Technical Focuses

To excel in throughput, manufacturing management primarily revolves around four key elements: three main pillars and a common foundation, as illustrated in Figure 1. This book focuses primarily on the technical and methodological aspects of these elements and their financial implications.

The three pillars include production management, maintenance management, and quality management. These major, cross-disciplinary technical topics significantly impact the operational throughput performance of manufacturing systems in various ways. They also present various opportunities and challenges for improvement, both individually and collectively. System and process designs, constituting the foundation, are essential for achieving manufacturing excellence.

However, other crucial factors that influence throughput improvement projects, such as teamwork, leadership, and communication, are not covered in this book. While equally vital for achieving excellence, they fall beyond the scope of this book.

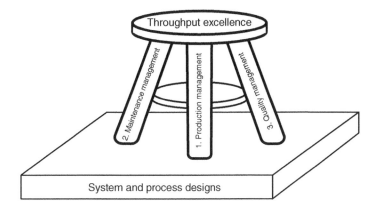

Figure 1   Main pillars and foundation for throughput excellence.

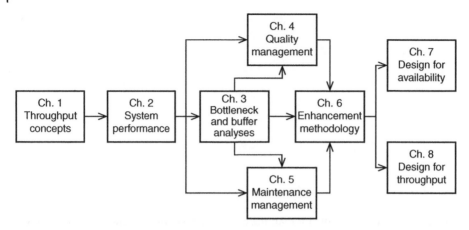

Figure 2 Contents and flow of this book.

The book is organized into eight chapters, encompassing the three pillars and the foundation of system throughput, as illustrated in Figure 2. Each chapter begins with general principles before delving into specific aspects and cases of manufacturing operations. While each chapter can be read and studied independently, it is recommended to follow the roadmap and cross-reference related chapters to acquire a better understanding of their interrelation and influence on manufacturing throughput performance. Armed with these principles and methods, you can cultivate a new mindset and habit of systematically pursuing manufacturing excellence.

Intended Audience

This book serves as both a textbook and a training manual designed for current and aspiring manufacturing professionals. It is well suited for graduate and senior undergraduate students in engineering, manufacturing, operational management, and related courses. In addition, it proves valuable for practicing manufacturing professionals looking to systematically enhance their knowledge and skills in improving operations on the production floor.

Adopting an operational management, continuous improvement, and systems design perspective, the book addresses the processes and challenges of enhancing production throughput. It provides essential principles, processes, and tools to achieve and sustain optimal throughput in complex system settings. However, its primary focus is on system performance and does not delve into some other aspects of manufacturing management.

This book aims to bridge the gap between the latest academic research and the practices of manufacturing practitioners by emphasizing applications over theory. It presents a unique view of operational excellence for manufacturing professionals, graduate students, and researchers. Drawing on hands-on experience in both industry and academia, the author integrates engineering principles, business perspectives, research work, and practical practices for the work on the production floor. While not a purely theoretical work, the book can serve as a valuable reference for academic researchers interested in further exploration.

Chapter Overview

As a comprehensive review of throughput management and enhancement in manufacturing operations, this book explores various related subjects across eight chapters.

Chapter 1: Throughput Concepts. This chapter introduces the fundamentals of throughput in manufacturing operations, covering its cost perspective, time analysis, system characteristics, basic operational states, and the standalone state concept.

Chapter 2: System Performance Metrics. This chapter explains how to evaluate operational performance using various KPIs from ISO 22400. The chapter examines OEE and its weighting, introduces the process of KPI selection, and explores the financial and accounting implications.

Chapter 3: Bottleneck Identification and Buffer Analysis. Covering the understanding, identification, analysis, and management of bottlenecks in manufacturing systems, this chapter explores buffer functionality, effect estimation, and status analysis for system throughput.

Chapter 4: Quality Management and Throughput. This chapter delves into quality definitions, quality management, and cost analysis. It explores the quality contribution to system throughput, including the analysis of quality issues in both serial and parallel systems, appraisal approaches, and improvement tools.

Chapter 5: Maintenance Management and Throughput. Introducing maintenance principles and strategies, this chapter addresses the impact of maintenance on throughput performance. It includes a comparison of maintenance strategies, reliability-centered maintenance, maintenance performance, and total maintenance cost analysis.

Chapter 6: Throughput Enhancement Methodology. This chapter outlines various approaches for problem-solving and continuous improvement, focusing on key characteristics and processes for throughput performance. It reviews throughput analysis and addresses throughput project management.

Chapter 7: Analysis and Design for Operational Availability. This chapter presents system design processes and approaches for ensuring the reliability of manufacturing systems by design. It proposes throughput-focused FMEA, examines the reliability of serial and parallel systems, and addresses availability reinforcement.

Chapter 8: System Design for Throughput Assurance. This final chapter reviews design strategies for throughput performance, including buffer planning, total cost analysis, capacity balance via cycle time design, value assessment, and computer simulation.

Book Features

This book presents the principles and methods of data-driven and bottleneck-focused approaches, emphasizing their applications to volume production manufacturing processes. While many examples are drawn from automotive manufacturing as a typical mass production setting, these approaches, with certain adjustments and adaptations, are applicable to other types of manufacturing processes, including low-volume and batch productions of discrete products.

To aid readers in comprehending and applying these principles and methods, the book incorporates a range of features crafted to facilitate learning and practical application. Each chapter concludes with 20 key takeaways, and each subsection, e.g., 2.4.3, has a brief summary for reference. Illustrated with 212 diagrams and 49 tables, this book succinctly demonstrates concepts and clarifies content through examples. Backed by about 270 cited sources, it provides a wide and rich exploration of throughput analysis, research, and improvement. This allows readers to uncover new opportunities for developing effective methods tailored to their specific situations.

Mastering the principles and methods of throughput improvement demands practical applications. To facilitate hands-on learning and in-depth understanding, each chapter includes analysis problems with solutions and 20 review questions for readers' exercise. The author, a seasoned manufacturing practitioner and researcher, also recommends engaging in a real throughput improvement project while reading this book or taking a related class.

The objective of this book is to equip readers with the knowledge and proven techniques necessary to attain throughput excellence and transform their production operations. Turn the page and delve into mastering those methods and techniques that can truly transform your production operations.

Acknowledgments

Industry and Academia Reviews

I express my sincere gratitude to Dr. Ziv Barlach (Chrysler, retired) and Carlos Zaniolo (Volvo Group) for their meticulous review and perceptive critique on the entire manuscript draft. I also appreciate the insightful feedback on some chapters and materials by senior professionals from various industries and academia, including Nick Deanes (APPLIED Adhesives), James Metzger (Emergent BioSolutions), Dr. Christopher Kluse (Bowling Green State University), Mike Rall (Dematic), Dr. David Tao (University of Michigan), Dr. Nasim Uddin (Global Automotive Management Council), and Dr. Jay Zhou (Ford, retired). Many others also provided valuable feedback.

I am grateful to the anonymous reviewers for their constructive feedback on this book's proposal and a preliminary sample chapter. Sincere thanks also go to Wiley's acquisition, project, editing, and publication teams for their contributions to the refined publication of this volume.

All Supports

I have been fortunate to collaborate with and learn from my experienced colleagues, mentors, and superiors in real-world manufacturing settings. I remain perpetually grateful to the professors who taught and advised me throughout my undergraduate and graduate studies. Their contributions to my education, industry career, and academic research have been invaluable and have built a solid foundation for this book. I also appreciate the authors whose excellent works I have referenced in the book.

I thank Eastern Michigan University for supporting the preparation of this book manuscript with a 2023 Sabbatical Leave Award. Finally, I express my heartfelt gratitude to my wife for her understanding and support in bringing this volume to fruition.

Your Feedback

Managing and improving manufacturing throughput benefit from a blend of art and science, calling for broad and diverse approaches. I have thoroughly considered them in this manuscript, drawing on my understanding, experience, and research in the field. I aspire for this book to serve as a valuable resource for those seeking to comprehend, apply, and advance manufacturing throughput practices.

I welcome and appreciate your insights, remarks, and feedback, which will contribute to the continuous enrichment of our knowledge and the refinement of this book. Please feel free to share your comments, critiques, and recommendations at htang2@emich.edu or htang369@yahoo.com. I will carefully review them for potential integration into future editions of this volume. I wish you immense success in your professional endeavors within the field of manufacturing.

January 2024

He (Herman) Tang
Ann Arbor, MI, USA

About the Companion Website

This book is accompanied by a companion website:

www.wiley.com/go/Tang/ManufacturingSystem

This website includes:
- Sample Syllabus
- Sample Instruction Schedule
- Solutions Manuals

1

Throughput Concepts

1.1 Introduction to Throughput

Throughput refers to "the amount of work, people, or things that a system deals with in a particular period" [Macmillan n.d.]. In the context of manufacturing, throughput denotes the quantity of products that can be produced within a period and the available resources.

Enhancing throughput is a central goal in manufacturing management, directly connected to financial performance and supported by various factors. These factors include daily management activities, methodologies, and technologies. Furthermore, these diverse factors can mutually influence each other in the pursuit of operational excellence.

Concerning the value creation of a business, operational management places significant emphasis on throughput in both manufacturing and nonmanufacturing sectors. Professionals widely acknowledge its importance, as exemplified by the statement, "throughput is king" [Miller 2021].

1.1.1 Role of Throughput on Business Success

1.1.1.1 Role of Manufacturing in Business

Business performance can be assessed from different perspectives and using various metrics, with key dimensions including financial health, customer satisfaction, and internal process effectiveness, as illustrated in Figure 1.1. At the core of internal processes for product-based businesses lies manufacturing operations.

Figure 1.1 highlights the interconnected nature of financial health, customer satisfaction, and internal processes. For instance, improving internal processes can boost productivity, positively affecting both financial health and customer satisfaction.

Manufacturing, as a central pillar for numerous companies, holds immense significance for overall business performance. Shortfalls in throughput targets lead to reduced production output, delayed deliveries, customer dissatisfaction, negative effects on the company's reputation, and diminished revenues. This book primarily emphasizes the effectiveness of internal processes within manufacturing systems, focusing specifically on productivity and throughput.

1.1.1.2 Financial Significance of Throughput

A company's operational profit can be expressed as the difference between the revenue generated by product sales and the expenses incurred from producing products and operating systems. Here

Manufacturing System Throughput Excellence: Analysis, Improvement, and Design, First Edition. Herman Tang.
© 2024 John Wiley & Sons, Inc. Published 2024 by John Wiley & Sons, Inc.
Companion website: www.wiley.com/go/Tang/ManufacturingSystem

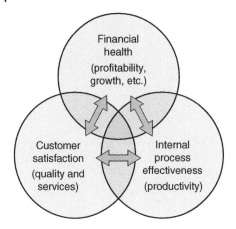

Figure 1.1 Main performance focuses of a manufacturing company.

is a simplified representation of the components contributing to profit:

Profit = Production rate × Unit sales

  – Direct costs (production materials and labor per unit × units produced)

  – Indirect production costs (utilities, other materials, indirect labor, and inventory, etc.)

  – Other indirect costs (overhead, general and administrative, etc.)

In this equation, the production rate significantly influences profitability by determining the output volume for the market. Thus, system throughput is an integral part of a business's financial success. Enhancing throughput aligns with sound business logic and can have a significant impact.

Indeed, manufacturing performance extends beyond throughput and encompasses a broader spectrum of factors. Elements such as customer satisfaction, employee engagement, market share, supplier relationships, and societal impact all contribute to the overall evaluation of manufacturing performance. While these aspects are not the primary focus of the book, their examination in conjunction with system throughput can yield valuable insights into their interplay. A holistic understanding of manufacturing operational performance can pave the way for more comprehensive and effective management and improvement strategies.

1.1.1.3 Work Focuses

A survey of 240 professionals from various manufacturing sectors across North America highlighted their main concerns in an increasingly competitive landscape marked by rising input costs and shrinking margins. These professionals identified efficiency and productivity as their top priorities [Alithya 2023].

Another survey of 153 US manufacturing executives indicated robust growth expectations within the industry [Kronos 2016]. The survey also identified the following 14 main operational challenges:

1. Improving internal production processes
2. Strengthening customer relationships
3. Finding enough people with the right skills and talent
4. Increasing labor productivity
5. Increasing demand responsiveness

6. Maximizing capacity and asset utilization
7. Meeting customer demands for product customization
8. Achieving annual cost reductions
9. Improving product and service quality
10. Responding to customer requests for quotes and proposals
11. Improving labor flexibility
12. Enhancing supply chain collaboration
13. Optimizing supply chain performance
14. Faster and more frequent new product releases and launches

Particularly related to manufacturing throughput, several challenges identified in the survey, such as 1, 4, 6, 8, and 9, are closely related to manufacturing throughput topics addressed in this book. For instance, Chapter 3 discusses how to identify and remove system bottlenecks; Chapter 6 presents methodologies for enhancing throughput; Chapters 7 and 8 offer proactive design considerations for improving production processes, aligning with improving internal production processes.

Regarding financial implications and cost reduction, Chapter 2 addresses throughput finance. Chapters 4 and 5 introduce the concept of the total cost of integrating throughput performance with quality and maintenance management, relating to challenge 8 – achieving annual cost reductions.

1.1.1.4 Throughput Management

Throughput management can be visualized as a pyramid-shaped business model with multiple levels (see Figure 1.2). This model is driven by a long-term growth vision. Manufacturing professionals align with the organizational mission (goals) and strategies, following relevant principles and using appropriate approaches to achieve improved performance. At its core, manufacturing throughput is propelled by the overarching business mission, strategy, and financial considerations.

In addition to system throughput, product quality is another fundamental performance aspect that impacts the overall system's performance. Both system throughput and product quality can be influenced by other factors, such as product design, manufacturing processes, operational complexity, and their interactions, as illustrated in Figure 1.3.

As financial health is the ultimate objective of any business, achieving excellence in throughput directly contributes to this bottom line, as mentioned earlier. Accounting and economic analysis can elucidate the correlation between throughput performance and financial outcomes. This crucial topic will receive comprehensive coverage in Chapter 2 of the book.

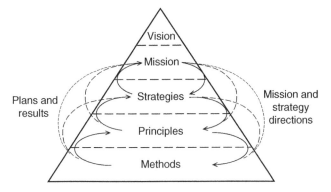

Figure 1.2 Business model for system throughput management.

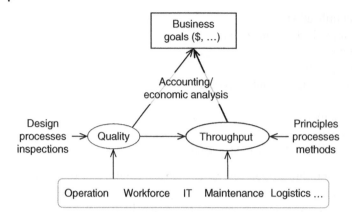

Figure 1.3 Manufacturing throughput management environment.

1.1.1.5 Throughput Work Collaboration

Effective system throughput relies on the collaboration among various departments involved in designing, operating, and supporting manufacturing operations. In this context, production management, maintenance management, and quality management emerge as three principal areas integral to throughput performance and enhancement, as illustrated in Figure 1.4. This book extensively covers these key areas, while not delving into other elements such as supply management and team management. Furthermore, Chapters 7 and 8 will explore in detail the significance of system design in shaping system capability and operational performance.

Managing and enhancing the performance of a manufacturing system is often considered both an art and a science, or perhaps a science-based art. Effective throughput management requires extensive knowledge, a diverse skill set, and substantial experience due to the system complexities arising from numerous variables and their intricate interactions.

Academic research in the realm of system throughput has a history dating back to the 1950s, with early examples such as Koenigsberg [1959]. Many analytical studies at that time focused on simple systems with only a few machines. Recently, computer simulation has gained prominence and will be explored in more detail later in this book.

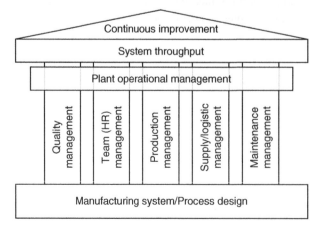

Figure 1.4 Pillars and foundation of manufacturing system throughput.

Subsection Summary: Manufacturing throughput plays a pivotal role in business, significantly impacting its financial performance and internal processes. Managing throughput improvement needs unique focuses and presents specific challenges.

1.1.2 Manufacturing Throughput Performance

1.1.2.1 Basics of Throughput

A manufacturing system transforms inputs, such as capital, materials, energy, machinery, labor, and information, into finished goods for customers. This transformation process includes a range of manufacturing functions, such as machining, assembly, and material handling, which may occur sequentially or concurrently. Operational management is concerned with the effectiveness of transformation processes, measuring productivity, throughput, and profitability, irrespective of process types and functions. Figure 1.5 illustrates how these terms relate to a manufacturing system.

The throughput rate (TR) of a system is a key metric used in operational management, defined as the number of product units produced within a specific period:

$$TR = \frac{Product\ units}{Period}$$

The unit of TR depends on production volume. For systems that produce large products like vehicles, TR is typically measured on an hourly basis, for example, 90 jobs per hour (JPH). Meanwhile, for small parts with high production volume, system throughput may be measured in pieces per minute.

TR has two basic types: design and measured. Design TR represents the target output or capability according to the system design. On the contrary, measured TR indicates system performance, largely related to operational management and measured on the production floor.

Throughput performance is typically assessed by expressing measured TR as a percentage of design TR. For instance, a measured TR at 95% of the design TR indicates that the operation fell short of the target by 5%, suggesting improvement potential for throughput performance.

In most of repetitive and continuous productions with a fixed design TR or process pace, total production output can also be used as a measurement of throughput or be presented as at both a rate and an amount.

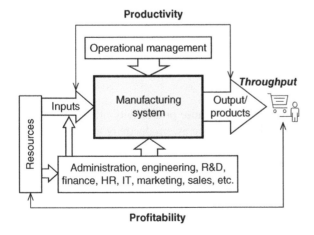

Figure 1.5 Systems view on manufacturing performance.

These concepts also apply to various nonmanufacturing sectors, such as retailing and financial services. In those sectors, throughput signifies the volume of material, service, or product that passes through the system and the final products delivered to customers. In healthcare settings such as hospitals, patient throughput can be assessed by measuring wait times and the length of stay.

1.1.2.2 Operational Availability

In managing a manufacturing system, terms like "availability" or "operational availability" are frequently used alongside TR. In many cases, throughput and availability are considered similar concerns and are related to factors such as process capability, product quality, and equipment maintainability.

The operational availability (A) of a system (or equipment, operation, etc.) is defined as:

$$A = \frac{\text{Actual production time}}{\text{Planned production time}} \, (\%)$$

In this equation, actual production time is the time spent producing products, excluding all nonworking time due to failures and other reasons. The planned production time represents the expected operating time.

For example, if a shift ran for 7 hours with a planned production of 7.5 hours, the system availability is,

$$A = \frac{\text{Actual production time}}{\text{Planned production time}} = \frac{7}{7.5} = 93.3\%$$

Organizations may have their own definitions. Toyota, for instance, considers starved time, equipment time, and work delay time when calculating the actual production time for the operational availability of a system [Sakai and Li 2020]:

$$A_{\text{toyota}} = \frac{\text{Planned production time} - \text{Starved time} - \text{Equipment time} - \text{Work delay time}}{\text{Planned production time}}$$

Here, starved time is the waiting time for parts from the upstream and other operations (discussed in Section 1.4); equipment time represents the equipment's downtime; and work delay time accounts for any additional time taken for the operations.

For example, if a shift was planned to run production for 7.5 hours, with a starved time of 0.3 hour, downtime of 0.2 hour, and work delay time of 0.1 hour, the corresponding availability would be:

$$A_{\text{toyota}} = \frac{7.5 - 0.3 - 0.2 - 0.1}{7.5} = \frac{6.9}{7.5} = 92.0\%.$$

The term "reliability" is often used to describe operational availability as the primary factor associated with availability. The reliability of a system (or workstation, work cell, equipment, etc.) will be discussed in detail in Chapters 2 and 4 of the book.

1.1.2.3 System Productivity

TR and productivity are sometimes used interchangeably because productivity often focuses on throughput and quantity. Typically associated with larger systems, productivity addresses individual or aggregate input and output, involving various resources and functions, such as labor, material, energy, and their combinations. A manufacturing system is deemed productive if the deployed resources and activities effectively add value to the products for customers. In the long run, high productivity leads to increased profitability.

As depicted in Figure 1.5, productivity is a ratio of input to output, calculated as:

$$\text{Productivity} = \frac{\text{Output}}{\text{Input}}$$

Here, the output represents the result of a process (measured by quantity, money, etc.) and the input represents the resources or effort put into that process (measured by money, time, etc.). The productivity ratio can be a dimensionless percentage or a value per unit, such as revenue per hour, per employee, or per piece of equipment.

For instance, in a vehicle assembly plant with 2200 employees producing 155,000 cars in a year (2290 working hours), labor productivity can be measured in several ways:

- Units per employee: $\dfrac{\text{Produced units}}{\text{Number of employees}} = \dfrac{155{,}000}{2200} = 70.45 \text{ units/employee}$

- Plant production rate: $\dfrac{\text{Produced units}}{\text{Production hours}} = \dfrac{155{,}000}{2290} = 67.69 \text{ units/hour}$

- Hours per unit: $\dfrac{\begin{array}{c}\text{Number of}\\\text{employees}\end{array} \times \begin{array}{c}\text{Production}\\\text{hours}\end{array}}{\text{Produced units}} = \dfrac{2290 \times 2200}{155{,}000} = 32.50 \text{ man-hours/unit}$

If the profits of the car models are known, the plant's financial productivity can be calculated at \$/hour. Moreover, productivity can be calculated at various levels within an organization, including specific processes, areas, equipment, and so on.

1.1.2.4 Considerations in Productivity

Improvement in productivity may be achieved by increasing output, using reduced resources, or employing a combination of both strategies. For instance, higher maintenance costs or added investments in new technology, treated as system input variables, can improve productivity. In such cases, the measurement of the productivity ratio becomes more important than measuring TR itself, as it helps evaluate the effects of input changes on system performance.

In some cases, especially in mass production, the primary inputs, such as direct labor and materials, remain consistent over a certain period. In such situations, operational performance can be measured solely by system throughput, without considering the variations in input. This implies that higher throughput yields more output with the same input of resources, such as labor and materials.

It is worth mentioning that continuously increasing labor productivity can be challenging. For instance, a study indicated that the labor productivity in the US fell by 2.8% between 2011 and 2021 after a significant increase between 2001 and 2011. Between Q4 of 2020 and Q4 of 2022, it declined by 0.4% [Clay 2023]. On the production floor, the challenge holds true for other types of productivity and throughput improvement.

Subsection Summary: Various aspects of throughput performance, including TR, availability, and productivity, form the foundation for understanding throughput performance and further analysis.

1.2 Discussion of Manufacturing Throughput

1.2.1 Characteristics of Manufacturing Throughput

1.2.1.1 Cost Perspective

Cost control, aimed at reducing both direct and indirect production costs, is an integral part of manufacturing management. Direct costs include expenses directly associated with production, including labor, parts, and materials. While indirect costs comprise expenses not directly tied to production, covering areas such as equipment, utilities, overhead, and support functions.

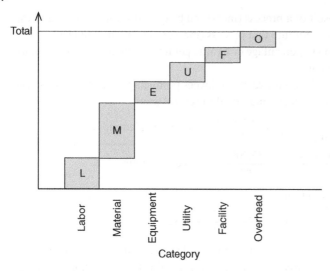

Figure 1.6 Main cost categories of manufacturing operations.

The total cost of a manufacturing operation within a given period can be presented as the sum of the following categories (refer to Figure 1.6):

$$C_{\text{total}} = C_{\text{labor}} + C_{\text{material}} + C_{\text{equipment}} + C_{\text{utility}} + C_{\text{facility}} + C_{\text{overhead}}$$

where

- C_{labor} represents direct labor costs, proportional to the number of production workers and front-line team leaders.
- C_{material} includes the costs of materials directly used in product manufacturing.
- $C_{\text{equipment}}$ represents equipment costs, based on depreciation of all machines, tools, and facilities (excluding the initial costs).
- C_{utility} accounts for utility costs, including expenses related to electricity, natural gas, as well as water for the system, etc. (some may be categorized as indirect costs).
- C_{facility} encompasses facility costs, including building depreciation, operations, maintenance, and related expenses.
- C_{overhead} refers to overhead costs, which are a predefined distribution of indirect costs covering items such as indirect materials, indirect labor, inspections, financial costs, and many others.

The relative sizes of the cost category blocks in Figure 1.6 are for illustration purposes, and the actual significance of each cost category varies depending on the types of manufacturing processes, system sizes, etc. For instance, energy-intensive processes, such as painting operations in an automotive assembly plant, incur high utility costs (C_{utility} in the equation). Processes involving extensive manual labor tend to incur significant direct labor costs (C_{labor}), particularly in countries with high wage levels.

It is important to note that the direct material cost (C_{material}) is proportional to the number of products produced. In contrast, other cost items (e.g., $C_{\text{equipment}}$, C_{utility}, and C_{overhead}) exhibit small variations in many cases within a given period. Direct labor costs increase with the quantity of products produced and the number of working hours.

Total cost can be expressed as a unit cost or cost per unit, serving as an indicator of operational effectiveness, productivity, and throughput performance. In many manufacturing scenarios, an objective is to reduce unit costs, with predetermined targets guided by historical data and the strategic goals of the production plant. Enhancing overall productivity can effectively lower unit costs, highlighting the value of throughput improvement.

1.2.1.2 Throughput and WIP

Within a manufacturing system, there exist many unfinished products, collectively known as work in process (WIP). While WIP does not generate value, it consumes capital through additional work, including tracking, storage, reprocessing, and inspection. Therefore, in alignment with the Lean principles and the pursuit of cost reduction, it is advisable to keep WIP as an operating expense at a minimal level.

In manufacturing, throughput time (TT) refers to the total time a product unit takes to traverse from the system's beginning to its end. TT serves as a tool for estimating the number of WIP units within the system, as illustrated in Figure 1.9 in the next subsection 1.2.2.

Little's Law, initially proven by John Little [1961] and subsequently examined by other researchers, establishes a relationship that remains unaffected by factors such as arrival process distribution, service distribution, service order, or any other considerations [Simchi-Levi and Trick 2011].

The long-term average WIP in a system can be calculated using Little's Law, given the known TT and TR:

$$WIP = TT \times TR$$

For instance, if the average TT for a system is 40 minutes (or $\frac{40}{60}$ hour), and the measured TR is 70 JPH, then the average of WIP units over an extended period would be:

$$WIP = \frac{40}{60} \times 70 = 46.7 \approx 47$$

When issues in the system impact throughput, such as system downtime or product defects, the actual TT may exceed the designated one, leading to a corresponding increase in the number of WIP units.

Elevated WIP levels lead to higher inventory costs, which may not be always considered in throughput studies. Reducing WIP levels is a strategic business move, albeit challenging, as efforts must be directed toward decreasing either TT, TR, or both. In certain situations, decreasing TT can be achieved through the implementation of Lean principles, workflow optimization, and the adoption of new technologies. Conversely, improving the measured TR to align with the design TR can also be an effective method for reducing WIP.

1.2.1.3 Continuous Improvement vs. Optimization

To enhance throughput performance on the production floor, a widespread practice is continuous improvement (CI). Although the terms "continuous improvement" and "optimization" are often used interchangeably in various cases, they differ in several aspects.

- *Conceptually*, CI involves an ongoing effort to incrementally enhance system performance. It is a routine and never-ending process, often recognized as both a mindset and a strategic approach. Breakthroughs can occasionally be achieved via CI efforts.

  In contrast, *optimization* involves the pursuit of the best possible outcome under a given situation, signifying a one-time effort to attain an optimal outcome. As a more academic term, optimization requires an in-depth study of mathematical modeling and computer programming. In some cases, optimization efforts may fail to produce solutions with certain algorithms and/or constraints.

- *Methodologically*, *CI* employs established approaches, such as PDCA and DMAIC (discussed in Chapter 6), conducted by field practitioners. It often integrates with other principles, such as Lean and Six Sigma, utilizing methods such as statistical analysis.

  Optimization encompasses specific methods aimed at maximizing or minimizing an objective function with specific mathematical models. Due to the technical complexity involved, optimization studies are typically conducted by well-trained engineers and researchers.

- *Process wise, CI* involves identifying opportunities for improving effectiveness/efficiency, enhancing quality, reducing waste. This process typically involves a manual interactive approach with multiple steps, including brainstorming, review, root cause analysis.

  In contrast, *optimization* aims to maximize or minimize a function output with input factors and constraints. It is typically conducted using a computer with optimization algorithms that automatically seek the best solution. However, the results must still be reviewed and interpreted for their feasibility in the real world.

- *Application wise*, CI is a universally applicable concept, implementable across various fields and widely used, especially in manufacturing.

  On the contrary, *optimization* is particularly suitable for engineering system design (discussed in Chapters 7 and 8), aiming to optimize resource utilization, achieve maximum performance, and potentially reach an ideal balance within theoretically designed systems.

Figure 1.7 depicts the processes of CI and optimization in a simplified two-dimensional world, acknowledging that both processes can involve more dimensions.

There is a wealth of literature available on both CI and optimization, with recent review papers by Skalli et al. [2022] and Cunha et al. [2023] for CI, and Yelles-Chaouche et al. [2021] and Renna et al. [2023] for optimization.

Integrating both methodologies for industrial applications is recommended, as it can offer complementary benefits. However, it can also be technically challenging due to the distinct nature of these approaches.

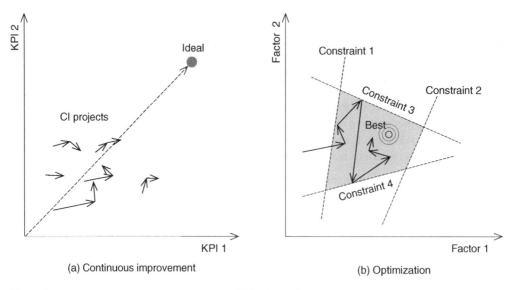

(a) Continuous improvement

(b) Optimization

Figure 1.7 (a) Continuous improvement versus (b) Optimization.

Subsection Summary: Enhancing manufacturing throughput involves cost control, WIP reduction, and CI. Little's Law establishes a link between TT and WIP. Integrating optimization and CI can optimize throughput excellence.

1.2.2 Time Analysis in Manufacturing Operations

1.2.2.1 Time Elements of Operations

Manufacturing operations do not always run perfectly because of unavailable equipment and blocked operations, among other factors. Understanding their time characteristics can aid in problem identification and solution.

The time elements during production can be recognized based on the state and task of an operation, as depicted in Figure 1.8 for their relationship:

- Planned time: Planned time is the total production time planned, excluding scheduled breaks, for example, 7.5 hour production time in an 8-hour shift, excluding 30 minutes of three breaks.
- Effective planned time: Effective planned time is typically calculated as available time + time loss due to some slow operations.
- Available time: Available time includes actual production time, blocked time, and starved time.
- Unavailable time: Unavailable time includes downtime resulting from failures and setup.
- Actual production time: Actual production time is the time of a system running to produce products.
- Blocked and starved time: Blocked and starved time is the waiting time due to external factors, to be discussed in Section 1.4.
- Setup time: Setup time is for support tasks, such as planned maintenance, adjustments, and tool changes.

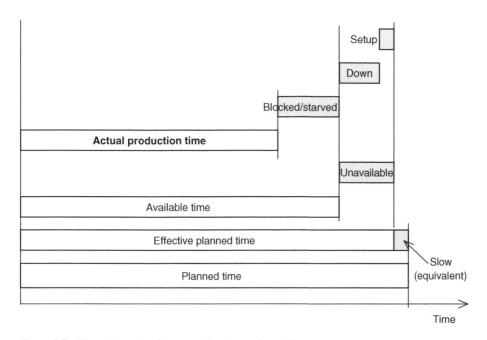

Figure 1.8 Time elements of an operation (or equipment).

These time elements and their relationships apply to various types of operations, such as equipment, workstations, assembly lines, and areas with multiple lines. Some elements may be broken into detailed items. For example, downtime may be due to equipment problems or quality issues. With clear definitions, categorizing actual data into these time elements can be straightforward.

Note that the quality of the products produced is not factored in here. If needed, the equivalent time of producing defective parts and/or reprocessing can be included. The influence of product quality is discussed in depth in Chapter 4.

Operational slowness is often minor and distributed throughout work time. Identifying and resolving slow operations require special attention. Slow operations often include minor stoppages. The classification of a stoppage as "minor" varies based on the specific operation. For instance, in the context of mass production, a stoppage lasting less than ten seconds might be considered minor downtime.

Moreover, with proper planning, setup, and maintenance times can be minimized or eliminated in the production time frame in most cases.

1.2.2.2 Definitions of Cycle Time (CT)

Cycle time (CT) is widely used in manufacturing and service industries as the fundamental operational pace. There are two common definitions of CT:

1. CT is the amount of time to complete one cycle of an operation, such as a workstation and system, from start to finish. Under this definition, the CT of a workstation and a line may differ (see Figure 1.9).
2. CT is the amount of time to complete work for a unit. This definition aligns with the CT meaning as a production rhythm or the time interval between consecutive products coming out at the end of a continuous operation. Under this definition, the CT of a workstation and a line are at the same level and are easy to compare.

The second definition facilitates a more consistent and comparable measure, especially when evaluating the throughput performance of a system at different levels. This definition is used throughout the book. Regardless of which CT definition is used, it is important to maintain consistency in throughput management to avoid confusion.

For throughput monitoring and improvement, manually measuring and verifying the CT of target operations is often necessary. Commonly, a stopwatch app on a smartphone is used for CT measurement. Due to the variation and interaction of processes, CT measurement should be repeated at least five times to obtain a representative mean value. A more effective method for CT measurement is videotaping using a camera app. Video recordings can be replayed as evidence in reviews of improvement opportunities.

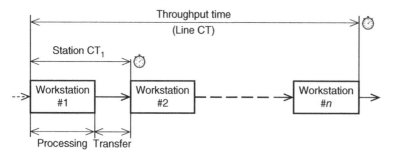

Figure 1.9 Workstation's cycle time and line's cycle/throughput time.

1.2.2.3 CT and Throughput

Similar to TR, CT has two fundamental types: design CT and measured CT. Design CT signifies the CT capability as per the design, whereas measured CT reflects the actual performance on the production floor.

In a repetitive process, when the CT is in seconds and the corresponding gross TR is in JPH, their theoretical relationship is expressed as:

$$TR_{gross} = \frac{3600}{CT\,(seconds)}\ (JPH)$$

Note that when CT and/or TR are in different units, the coefficient "3600" in the equation needs to be changed accordingly.

This equation demonstrates that TR and CT have a reciprocal relationship, as illustrated in Figure 1.10a. When the CT of an operation deviates from the CT, the TR changes inversely. For instance, if CT increases (slower) by 3%, the corresponding TR will decrease by approximately 3%. Figure 1.10b provides a close-up view of the region around CT = 60 seconds in Figure 1.10a.

In the formula of TR and CT, gross TR is based on the assumptions of perfect operational availability (or reliability), no interactions (i.e., starvation and blockage) between operations, and no variability. However, these assumptions do not hold true and are addressed in the later chapters. Given these limitations, the calculations based on this equation serve only as rough estimates. Considering the actual situation and relevant variables leads to the concepts of net TR, standalone TR, and measured TR, which are discussed in depth in subsection 7.2.1.

In a batch process, CT can be measured either per batch or per piece on average. For example, in a machining process that can cut 20 pieces within 15 minutes as a batch, the average CT of the process would be:

$$CT = \frac{15 \times 60\,(seconds)}{20\,pieces} = 45\ seconds\ (per\ piece)$$

For a repetitive process, internal transfer time within a system is typically included in CT (refer to Figure 1.9). For a batch process, transfer time can be averaged on a per-piece or per-batch basis. If a transfer function is external and/or not related to the start time of the next operation, it may be counted separately.

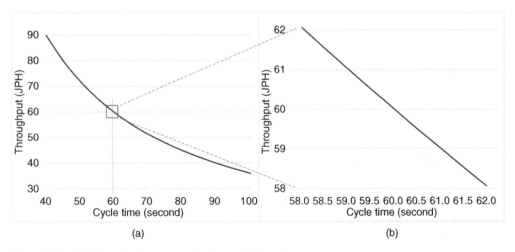

Figure 1.10 Relationship between cycle time and throughput rate.

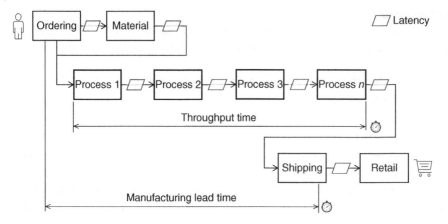

Figure 1.11 Throughput time and lead time of a manufacturing system.

1.2.2.4   Lead Time and Throughput

As mentioned in an early subsection 1.2.1.2, TT is the time span from start to finish for a product, as shown in Figure 1.9. TT is an important performance indicator, as it constitutes the major portion of the lead time to customers, showing system efficiency.

TT is primarily determined by product and process designs, especially in repetitive processes and certain batch processes. In TT, queue time (waiting time in the system) and setup time vary, which primarily depends on operational management and execution. Such elements of TT, including queue time, setup time, process flow gaps, and stops, do not add value to customers. Therefore, they should be focused on reducing TT for throughput management.

Another related term is manufacturing lead time, representing the entire time between receiving a customer order and delivering the product, as shown in Figure 1.11. In the figure, "latency" refers to the time spent on transfer, inventory, and other related activities. They can also be considered non-value-added to customers.

Illustrated in Figure 1.11, the TT of a manufacturing system is the core element of lead time, along with additional nonproduction latency time spent. As a metric, TT measures the internal efficiency of manufacturing processes, while lead time reflects the overall responsiveness to customer orders. Manufacturing professionals sometimes use the two terms interchangeably when focusing on production operations. Understanding and reducing them can enhance manufacturing system performance and benefit customer satisfaction.

Subsection Summary: Time perspectives, including time elements, CT, and the reciprocal relationship between CT and TR, offer effective lenses for analyzing manufacturing operations. Lead time and TT are also crucial measures of manufacturing efficiency.

1.3   Characteristics of Manufacturing Systems

1.3.1   Overview of Manufacturing Process Types

1.3.1.1   Four Major Types of Processes

Manufacturing processes can be categorized into four main types in terms of production volume and product variety (see Figure 1.12):

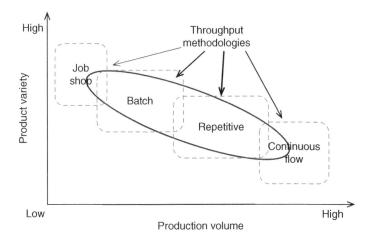

Figure 1.12 Characteristics of four types of manufacturing processes.

- Job shop process: In a job shop process, a single or small lot size of a customer's product is produced. The products and their manufacturing process are typically unique and not fully standardized. Setup conditions and process steps can vary. Examples of job shop processes include a large machine assembly and work in a repair shop, as well as services provided in hospitals and sit-down restaurants.
- Batch process: In a batch production, product units are grouped together with common process steps and tools. Products are typically manufactured in small batches of several units. Examples of batch production include biotech product manufacturing, food processing, sheet metal stamping for vehicle assembly, and fast-food restaurants. Depending on production volume and other factors, batch production can be either synchronous or asynchronous.
- Repetitive process: A repetitive process, also referred to as a discrete-product or transfer line process, entails the sequential processing of products for an extended duration. The manufacturing activities are repetitive for process steps. Repetitive processes often use a synchronous or one-piece-flow approach. An example is vehicle assembly, where the scale of plants can range from midsize to mega-sized facilities.
- Continuous process: A continuous process, also known as process manufacturing, operates continuously around the clock, maintaining a constant flow of materials or products. The process industry is typically characterized by large-scale operations and at high-automation levels. Industries that utilize continuous flow systems include oil refining, chemical processing, pharmaceuticals, and the food industry.

Throughput management and improvement is a shared concern across all four types with different focuses due to distinct characteristics. The methodologies of throughput improvement discussed in this book are most suitable for repetitive processes, easily adaptable to batch and continuous processes, a useful reference to job shop processes, refer to Figure 1.12. More discussion follows in the next subsection 1.3.1.2.

1.3.1.2 Comparison of Process Characteristics
A comparison of the characteristics of these four types of processes enhances our understanding of their differences, providing valuable insights for operational management and throughput improvement. Table 1.1 succinctly summarizes these characteristics. It is important to note that these characteristics are relative, with notable overlaps between process modes in many cases.

Table 1.1 General comparison of four types of manufacturing processes.

Attribute	Job Shop	Batch	Repetitive	Continuous
Cost per unit	High	Moderate	Low	Very low
Equipment	For general purposes	Some for special purposes	Most for special purposes	For special purposes
Worker skills	High	Moderate	Low	Various
Automation	None/low	Medium/high	High	Fully
Scheduling	Complex	Moderately complex	Routine	Routine
Main throughput KPI	TT	TT and TR	TR and quality	Cost and quality

With a shared concern on system throughput performance, there may be distinct aspects and key performance indicators (KPIs), as outlined in the bottom row of Table 1.1. These differences highlight the need for customized throughput management approaches and focuses for each process type. For example, in a batch process, due to the product variety from process, material, and batch sizes, etc., operational management typically addresses TT and TR as main KPIs. Conversely, due to high production volume in a continuous process, KPIs tend to be cost and quality.

There are other KPIs. For example, in batch and repetitive processes, two basic scenarios exist: assemble-to-stock (ATS) and assemble-to-order (ATO). In ATS, inventory availability is an important performance metric to support throughput. In ATO operations, delivery speed and reliability are crucial KPIs. Further discussion on KPIs is provided in Chapter 2.

Following market demands, a manufacturing system operating in a repetitive mode may need to produce more products with distinctive designs. This can prompt a transition from a repetitive mode to a batch mode in the manufacturing system. In such cases, in addition to the system's flexible capability to swiftly adapt its process mode, it is important to adjust throughput management to maintain effectiveness. Manufacturing flexibility is discussed in Chapter 8.

Subsection Summary: The throughput approaches and methods presented are readily applicable to batch and repetitive processes, requiring some adjustments for the job shop and continuous processes. The principles of throughput management can benefit all four types of manufacturing processes.

1.3.2 Settings of Manufacturing System

1.3.2.1 Systems View and Configuration

Manufacturing is a physical process that involves the conversion or transformation process of materials into finished products. As depicted in Figure 1.13, a manufacturing system comprises machinery, processes, methods, and personnel working together to convert raw materials into customer products. The goal of a manufacturing system is to produce products that meet customer expectations in functionality while minimizing resource use, costs, and delivery time.

This systems view focuses on understanding a system as a whole, considering the contributions and interactions of individual parts to the overall functionality and output of a system, rather than focusing on their individual functions. The understanding of systems views on manufacturing establishes a good foundation to discuss the configurations and composition of manufacturing systems.

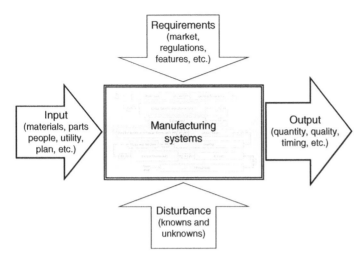

Figure 1.13 Conversion view of a manufacturing system.

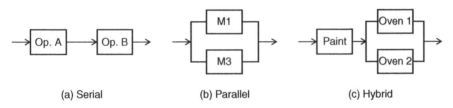

(a) Serial   (b) Parallel   (c) Hybrid

Figure 1.14 Basic configurations of manufacturing systems.

Manufacturing systems and their process flows can be categorized into three basic styles of system layout based on their operational sequences. This is depicted in Figure 1.14.

A serial configuration is prevalent in manufacturing systems, which accounts for most mass production, such as vehicle assembly. Furthermore, despite varying local configurations, the overall flow and layout of a complex manufacturing system are typically in serial.

A parallel configuration is commonly employed in situations involving slower operations and/or high-volume production. For instance, if a single machine tool can support half of the production volume required by the market, two machine tools can be designed in parallel to meet the production volume.

A hybrid configuration, combining serial and parallel elements, is common in large systems to balance process paces. For example, two or three parallel curing ovens are designed for a painting system, because the curing process takes much more time than the painting process.

Local branches and loops often exist within a manufacturing system to accommodate functional requirements, including repair, reprocess, scrap pull-out, and bypass. Two examples of such processes are shown in Figure 1.15. These local arrangements add to the complexity of the manufacturing system, which may have a significant impact on manufacturing throughput.

There are other layout configurations in manufacturing systems, which will be discussed in the later chapters. During bottleneck analysis, there are instances when hybrid and complex configurations may require decomposition into simpler serial and parallel elements, a topic that will be further discussed in Chapter 6.

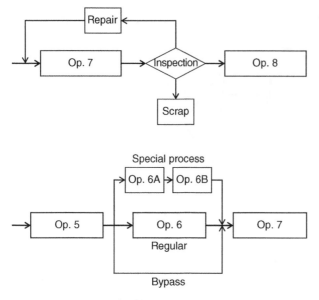

Figure 1.15 Local configuration examples in manufacturing systems.

1.3.2.2 System Composition

Manufacturing systems are complex networks that must be examined holistically using a systems-view approach. Recognizing the configurations and composition of the entire system is essential for analyzing and improving throughput performance.

In addition to system configurations, the systems perspective can be applied to the composition of a manufacturing system. System composition analysis is vital for addressing throughput problems at multiple levels. The system structure can be presented hierarchically, akin to an organizational chart, guiding an in-depth investigation. Figure 1.16 illustrates a manufacturing system in a simplified cascade view.

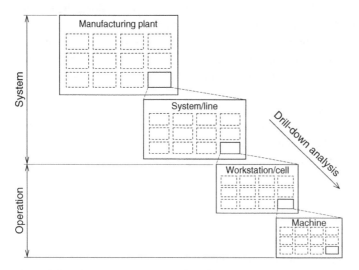

Figure 1.16 Cascade view of operations in a manufacturing system.

A comprehensive systems view is crucial for effective throughput improvement. Throughput problems and issues are often identified at a system level. Addressing and solving throughput issues in a system, bottleneck identification and root cause analysis involve analyzing multiple levels, drilling down to a specific machine or process. A solid understanding of the structure, flow, and relationships within a manufacturing system is essential to find root causes and resolve the problems. This process is discussed in more detail in Chapter 6.

The systems view of a manufacturing system extends beyond recognizing its structural complexity. It involves understanding the system as a network of interrelated subsystems, where the operational states and functions of each subsystem are not independent. The effects of subsystem interactions over time on system behaviors must be examined, in addition to identifying problematic issues.

For example, a problem occurring in one subsystem may have its root causes originating from other subsystems. Performance indicators, such as OEE (discussed in Chapter 2), can reflect a complex interplay of multiple factors. Without adopting a systems-view approach, throughput improvement is unlikely to be fully effective. Furthermore, improving an individual subsystem may or may not result in an overall performance improvement for the entire system, a critical point discussed in depth in Chapter 3.

Subsection Summary: A holistic systems view of manufacturing systems recognizes their functionality and complexity, considering both various configurations and compositions, enabling a better understanding for achieving throughput excellence.

1.3.3 Additional Throughput Considerations

1.3.3.1 Throughput Performance Pillars

To effectively manage throughput in a manufacturing system, it is essential to consider various factors. These factors can be grouped into three major categories or pillars, along with other miscellaneous factors:

1. System design
2. Production planning and control
3. CI on the floor
4. Other (miscellaneous) factors

Each pillar plays a unique supporting role, influencing throughput performance and presenting opportunities for improvement. Beginning with system design, it establishes a foundation for system throughput capacity. Engineering design professionals should address throughput during the design phases, a topic that will be discussed in-depth in Chapters 7 and 8.

The next pillar is production planning and control, which plays a crucial role in ensuring the smooth flow of operations and meeting production demands efficiently. It is an integral part of operational management, involving scheduling activities and considerations to allocate machinery, production processes, human resources, and raw materials. The control aspect of the pillar is for any necessary adjustments and corrective actions. While this book does not focus on this subject, it acknowledges its importance.

CI is a main activity of throughput management and a central theme of this book. After this introductory chapter, Chapters 2–5 will delve into the respective aspects of throughput management, and Chapter 6 will provide a more detailed introduction to the approaches and processes for throughput CI.

These factors are all important to any type of manufacturing operation, while their significance may vary depending on the specific context, which will be addressed in the next subsection 1.3.3.2.

In addition to the three main pillars, there are other or miscellaneous factors. They include numerous known and unknown ones, such as consistent changes and variations, on the production floor. Some common factors are discussed in the related throughput aspects in the corresponding chapters.

1.3.3.2 Throughput Improvement Potential

Associated with different types of manufacturing processes, the three main pillars exhibit varying potentials for improving throughput performance. Figure 1.17 illustrates the different significance of these pillars for discussion purposes. For example, in a repetitive process, the three pillars (system design, production planning and control, and CI) contribute approximately 45%, 20%, and 25% to the total potential, respectively. The actual relative improvement potentials vary depending on the specific manufacturing system and its context.

In the figure, CI presents distinct opportunities to enhance system throughput on the production floor (further discussed in subsection 6.4.3 in Chapter 6). For instance, the room for CI may be relatively smaller for continuous processes, as they are largely determined by system design. In addition, CI has limited room for improvement as some potentials are more related to other factors, such as system design or production planning and control.

The improvement potential of each pillar is rarely studied, and the limits of improvement are not extensively explored either. However, many experienced professionals understand that as time progresses, improving throughput through CI alone may plateau and become increasingly challenging without making major system or process design changes. Therefore, operational management should consider multiple pathways for improving throughput, frequently assess the effectiveness of their CI efforts, and closely cooperate with the engineering design teams.

1.3.3.3 Issues in Batch Process

Batch processes can handle a wide range of products, but their throughput depends more on production planning and control. This is because products may require unique process routes and durations. Figure 1.18 provides an illustrative example of batch production for multiple products with different processes across eight workstations within the same system.

In addition, tooling and fixture changes can occur often when switching between different products. Measuring and improving system throughput in such systems can be more complex due to the variety of process flows and timings.

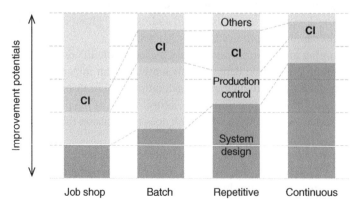

Figure 1.17 Relative potentials of system throughput improvement.

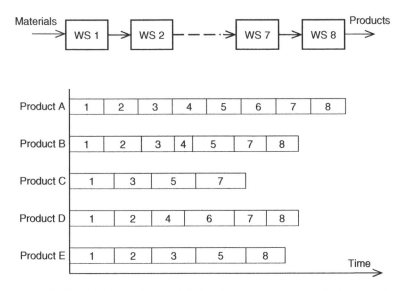

Figure 1.18 Multiple products and different processes in a manufacturing system.

Production planning and control plays a primary role in bath and job shop processes. It considers multiple factors, such as batch size, frequency, and process time, to optimize operational throughput and process utilization. When dealing with known processes, batch size is often a key consideration for resource utilization optimization (see Shafeek and Marsudi [2015] as an example).

Production planning and control for batch processes also considers factors such as order priority and WIP. A simple scheduling approach is First In, First Out (FIFO). Other scheduling approaches may aim for the shortest processing time, constant WIP (CONWIP), batch processing based on process similarity [Olaitan et al. 2017], or re-sequencing product mix considering part and process constraints [Fradkin et al. 2017].

To effectively manage and improve system throughput, a holistic approach across system design, production planning and control, and CI can reach its full potential (refer to Figure 1.17). The importance of each pillar needs further studies to address the special needs of a batch process.

Subsection Summary: Enhancing system throughput needs to address the following three pillars: system design, production planning and control, and CI. System design sets the capability, planning and control manages resources, and CI maximizes production potential. Different processes, like batch processes, pose distinct considerations.

1.4 Operational States of Production

1.4.1 Operational States of Systems

1.4.1.1 Nonworking States

To analyze system throughput levels, it is essential to know the operational states of a system. For throughput management, there are four types of stoppage or nonworking states:

1. Downtime due to the system's own failures (called faulted)
2. Starvation from parts by its upstream system (called starved)

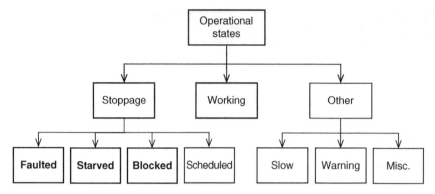

Figure 1.19 Operational states of manufacturing systems.

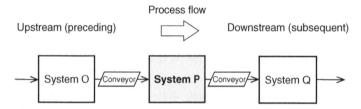

Figure 1.20 Three production systems in a serial configuration.

3. Blockage by its downstream system (called blocked)
4. Scheduled stoppage, such as for maintenance and setup

Figure 1.19 summarizes operational states for a manufacturing system, emphasizing four types of stoppage.

For instance, Figure 1.20 illustrates three connected systems using conveyors serially. If System P does not receive parts from its upstream System O on time, it is starved by System O and stops operating. Similarly, if System P cannot send the finished jobs to its downstream System Q, it is blocked by System Q and stops. In both cases, System P nonwork is not at fault but caused by *external* factors.

Furthermore, due to countless variations in operations, a system rarely achieves perfect synchronization inside even if it is designed to operate smoothly. Consequently, there are starvation and blockage situations between workstations within a production line. These internal starvation and blockage may be considered normal as long as their time remains minimal, typically less than 2% of the design CT. More discussion on internal interactions is in Chapter 8.

There can be also situations known as scheduled stoppages. Examples include maintenance scheduled during production and setup required for batch production. Even if a production planning and control department attempts to schedule them into nonoperational time, some tasks, such as equipment adjustments, may be unavoidable during production.

1.4.1.2 Discussion of Operational States

In the last subsection (Figure 1.19), the operational states of a system are discussed. For each type of state, there can be several substrates based on reasons. Table 1.2 lists some typical subtypes of operational states in mass production as an example.

Table 1.2 Operational states and types and their internal/external reasons.

Reporting priority	State	Type (sub-state)	Internal reason	External reason
100	Normal working	In design cycle	N/A	N/A
9	Scheduled stoppage	Scheduled break time		x
		Setup and changeover		x
		Scheduled maintenance		x
1	Stoppage (due to *Internal* issues)	Equipment fault	x	
		Computer/control fault	x	
		Safety fault	x	
		Quality fault	x	
		Manual production stop	x	
		Internal blockage	x	
		Internal starvation	x	
2	Stoppage (due to *External* factors)	Blockage (waiting)		x
		Starvation (waiting)		x
		No material/part		x
		Defective material/part		x
		Network/communication		x
3	Slowness	Automation slow cycle	x	
		Manual operation over the cycle	x	
		Minor stoppage	x	
		Quality alert	x	x
4	Warning	Low material alert		x
		Assistance required	x	x
		Manual intervention		x
5	Miscellaneous items	Nonscheduled production		x
		Tryout or test		x
		Idle		x
		Other/undefined	?	?

Subtypes, such as scheduled maintenance, can be further categorized based on activities, for instance, inspection, adjustment, calibration, cleaning, among others. In addition, there are warning types that serve as alerts before an actual failure occurs. The classification of categories and subcategories depends on the system's specificities and the process's characteristics.

Identifying operational state types enables effective system throughput management with quick responses and efficient problem-solving. Automated systems typically feature real-time reporting and data analysis, allowing operational management to promptly assess situations and respond.

1.4.1.3 Additional Thoughts on Operational States

Defining operational states for monitoring and tracking purposes is a critical task in system design. To enhance throughput performance and effectively address problems, engineering design professionals should have a comprehensive understanding of the possible internal and external reasons for each state. The right two columns, i.e., "Internal reason" and "External reason" of Table 1.2, offer the identification of internal/external reasons for the states, as a reference example.

Defining these subcategories of operational states is an ongoing effort when developing complex manufacturing systems. Thus, it is essential to include an "other/undefined" category in a state list to account for unknown or unclear situations that may arise beyond the defined items. Once the root causes of a specific situation are fully understood, it can be categorized into an existing state or defined as a new type of state. Continuously refining the state reporting enables accurate state identifications for operational management.

Monitoring screens typically display one system state at a time. However, sometimes multiple states may occur concurrently, such as equipment failure and blocking happening simultaneously. In such cases, the computer system needs to prioritize which state to display and report. Therefore, it is necessary to determine the reporting priority for each state, ensuring that the computer system displays the highest-priority state when multiple states occur. In Table 1.2, the left column, "Reporting priority," shows an example of the reporting priority, where "1" is the highest.

Defining the reporting priority of operational states, the stoppages because of the internal issues should receive the highest priority, marked as "1" in the table. Conversely, a normal working state should be given the lowest priority, such as "100," ensuring that any type of abnormal state can be reported without being concealed. Likewise, setting individual priorities for subcategories may be needed. Reporting priority settings are determined by operations management.

Subsection Summary: Identifying and interpreting various operational states, particularly nonworking states due to internal and external influences, is essential for assessing system status. A deeper discussion of the nonworking states follows in the next subsection 1.4.2.

1.4.2 Applications of Standalone State

1.4.2.1 Standalone State Concept

As discussed earlier, when a system is not working, the reasons can be either internal or external causes. External factors involve upstream or downstream functions or states that exist independent of the system. When problem-solving focusing on the system itself, it is essential to identify and exclude the external factors in the analysis for system throughput improvement.

A standalone state describes a situation in which the system operates independently of all other systems and external factors. The standalone concept for manufacturing production was introduced over 30 years ago [Hopp and Simon 1993] and has since been theoretically studied [Li et al. 2013, Aboutaleb et al. 2017].

Understanding the standalone concept is crucial because improvement efforts should be directed toward identifying and addressing the system's internal root causes rather than external factors. This standalone approach ensures the accurate identification and resolution of throughput issues. Mixing internal and external root causes of throughput issues can mislead problem-solving efforts and reduce improvement effectiveness.

Apart from the influences of upstream and downstream systems, production part logistics – supporting manufacturing operations with parts and materials – is another common cause of system starvation. Logistics issues, such as delays and misplacement, can result in system starvation and directly impact system throughput. Fortunately, these logistics issues are often visible and not technically challenging to identify and rectify.

1.4.2.2 Standalone Availability

To accurately evaluate a system's true performance, capability metrics are needed that exclude external factors. Standalone availability (A_{sa}) achieves this by considering only the system's internal failures and faulted downtime. A_{sa} is calculated using the following formula:

$$A_{sa} = \frac{\text{Actual production time}}{\text{Planned production time} - \textbf{Starved time} - \textbf{Blocked time}}$$

Comparing this to the operational availability (A) discussed in subsection 1.1.2.2, A_{sa} is higher because starved time and blocked time are normally present and deduced from the planned production time in the equation. Table 1.3 provides an example to demonstrate the difference between A and A_{sa}. Figure 1.21 illustrates the four operational states of this example.

The operational availability and standalone availability can be calculated as follows:

$$A = \frac{6.85}{7.5} = 91.3\%$$

$$A_{sa} = \frac{6.85}{7.5 - 0.15 - 0.3} = 97.2\%$$

Following the same concept, an alternative calculation of A_{sa} is:

$$A_{sa} = \frac{\text{Actual production time} + \textbf{Starved time} + \textbf{Blocked time}}{\text{Planned production time}}$$

Table 1.3 Example of a system's operational states.

State	Time	
Scheduled	7.5 hours	100%
Operational	6.85	91.3%
Stoppage:	0.65	8.7%
Faulted	0.2	2.7%
Staved	0.15	2.0%
Blocked	0.3	4.0%

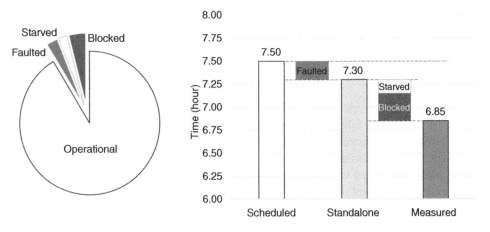

Figure 1.21 Example of four states of manufacturing operations.

By employing this alternative equation, the A_{sa} of the above example is calculated to be 97.3%, a slight deviation from the previous calculation of 97.2%. Either equation can be utilized, provided consistency is maintained with one method.

In industry practice, there are variations in the treatment of starvation and blockage when calculating standalone metrics. For example, Toyota includes starvation into operational availability [Sakai and Li 2020]. As a result, the availability and standalone availability across different companies may not be directly comparable due to the differing considerations.

1.4.2.3 Standalone TR

In addition to applying to operational availability, the standalone concept can apply to TR and OEE (to be discussed in Chapter 2) as system throughput KPIs, to exclude external influences on these metrics.

The standalone TR, denoted as TR_{sa}, can be defined and calculated based on the known values of A_{sa}, A, and measured TR:

$$TR_{sa} = \frac{A_{sa}}{A} \times TR$$

In this equation, the key is the ratio $\frac{A_{sa}}{A}$:

$$\frac{A_{sa}}{A} = \frac{\dfrac{\text{Actual production time}}{\text{Planned production time} - \textbf{Starved time} - \textbf{Blocked time}}}{\dfrac{\text{Actual production time}}{\text{Planned production time}}}$$

$$= \frac{\text{Planned production time}}{\text{Planned production time} - \textbf{Starved time} - \textbf{Blocked time}}$$

According to the equation, if a manufacturing system experiences no starved time and blocked time in a period, the system exhibits a ratio of $\frac{A_{sa}}{A} = 1$ or $A_{sa} = A$. Consequently, $TR_{sa} = TR$. When starved time and/or blocked time are present, the ratio $\frac{A_{sa}}{A} > 1$. That implies that $TR_{sa} > TR$. Using the data of the above example,

$$\frac{A_{sa}}{A} = \frac{97.2\%}{91.3\%} = 1.065$$

Based on the standalone concept, both A_{sa} and TR_{sa} reflect the true throughput performance of a manufacturing system but are from different perspectives. In industry practice, both metrics are often used at the same time.

1.4.2.4 Discussion Examples

Figure 1.22 illustrates an example of applying the standalone concept. In this example, a vehicle assembly plant consists of the following three shops: body frame, body paint, and general assembly, arranged in a serial setting and operating independently. Their design TRs are 80, 79, and 78 JPH, respectively. The figure displays each shop's throughput performance in a week with TR, standalone TR, itemized throughput losses due to faulted, starved, and blocked states.

From Figure 1.22, it is evident that the body frame shop had the lowest standalone TR and the longest faulted downtime, contributing to the starved time of both the body paint shop and general assembly shop. Thus, it is recommended that the body frame shop be identified as the first target for throughput improvement in the plant.

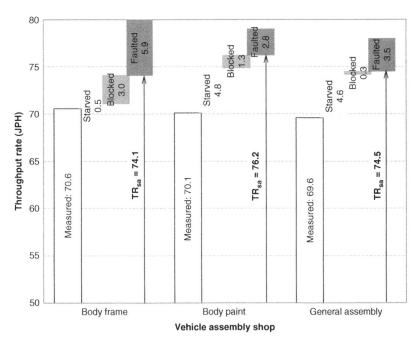

Figure 1.22 Example of throughput rate elements of three manufacturing systems.

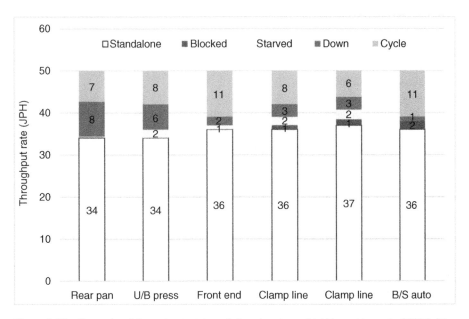

Figure 1.23 Example of throughput status of six subsystems (Valdés and Leoncio 2003/with permission of Massachusetts Institute of Technology).

Here is another example of using standalone throughput metrics [Valdés and Leoncio 2003]. Figure 1.23 illustrates a system with six subsystems, and the TRs of each subsystem are categorized as standalone, down, starved, or blocked. In this case, the authors referred to the unknown throughput loss as a "cycle" at the time. The rear pan area has the lowest standalone TR, making it the first target for throughput improvement.

Subsection Summary: The standalone concept, applied to availability and TR, improving their accuracy in reflecting true system performance and offering meaningful insights for throughput management. More discussion and applications continue in the later chapters.

Chapter Summary

1.1. Introduction to Throughput

1. Manufacturing system productivity, or throughput, is a crucial indicator of internal process effectiveness for overall business success.
2. Top manufacturing priorities include efficiency, productivity, addressing challenges in production processes, and maximizing capacity utilization.
3. Throughput management integrates system throughput, product quality, and financial outcomes, requiring cross-functional collaboration for long-term business growth.
4. Throughput, the transformation of inputs into goods, can be measured by TR and Operational Availability (A), etc.
5. System productivity involves enhancing output, resource utilization, and adapting to input changes.

1.2. Discussion of Manufacturing Throughput

6. Total cost breakdown includes labor, material, equipment, utilities, overhead, facility costs, etc., often measured as a unit cost.
7. Little's Law links TT and design TR.
8. CI and optimization differ conceptually, methodologically, and in application. The work on the production floor focuses on CI.
9. CT, a fundamental operational pace, is recommended as the time to complete a single unit of work for consistency.
10. CT and TR have a reciprocal relationship in an assumed perfect operational scenario.
11. TT and lead time are key indicators of manufacturing efficiency and customer responsiveness.

1.3. Characteristics of Manufacturing Systems

12. Manufacturing systems and processes are classified into the following four types: job shop, batch, repetitive, and continuous, with different considerations and influencing factors for throughput management.
13. A systems view analyzes the entire manufacturing system, considering configurations and composition.
14. System layouts include serial, parallel, and hybrid configurations, adding complexity with branches.
15. System design, production planning, and CI influence throughput, each with unique opportunities.

16. Batch size, frequency, order priority, and WIP require a holistic approach for effective throughput management.

1.4. Operational States of Production

17. Understanding nonworking states (downtime, starvation, blockage, scheduled nonproduction) is crucial for analyzing system throughput.
18. Categorizing and prioritizing reporting operational states enables effective throughput management and problem-solving.
19. Standalone state focuses on a system operating independently of external factors, pinpointing internal root causes accurately.
20. Standalone state can be applied to metrics, such as A_{sa} and TR_{sa}.

Exercise and Review

Exercises

1.1 A shift has 7.5 hours of planned production time. The actual production time in a shift was 6.8 hours. The shift experienced 0.15 hour of starved time, 0.25 hour of blocked time, 0.2 hour of equipment downtime, and 0.1 hour of work delay time. Calculate the operational availability and Toyota's formula availability for this shift.

1.2 A manufacturing system employs 320 people. Over one week, it utilized 13,400 labor hours and produced 4800 units. Calculate the productivity per employee and per hour for this week.

1.3 In one week, a shop incurred a direct labor cost of $100k, material cost of $150k, equipment cost of $45k, utility cost of $7k, and overhead cost of $13k. Calculate the percentage of direct labor cost in the total cost.

1.4 An operation has a known throughput time of 4.5 hours and a throughput rate of 50 jobs per hour. Calculate the average WIP level for this operation in the long run based on Little's Law.

1.5 A production line has eight workstations with cycle times of 58, 56, 57, 60, 59, 56, 55, and 59 seconds, respectively. Estimate the long-run average WIP for this production line.

1.6 If a process has a cycle time of 48 seconds, estimate the corresponding theoretical throughput rate in JPH.

1.7 Based on the data from Exercise 1.1 and known production output of 480 units, calculate the standalone availability (A_{sa}) and standalone throughput rate (TR_{sa}).

Review Questions

(The chapter covers these topics. For further discussion, it is recommended to seek additional information and examples. Diverse perspectives are encouraged.)

1.1 Discuss the relationship between internal process (operations or production) effectiveness and financial health, providing an example.

1.2 Address one of the top 14 manufacturing challenges mentioned and offer a solution based on a particular case.

1.3 Review how product quality directly and indirectly contributes to manufacturing throughput performance, providing an example.

1.4 Explore the similarities and differences between operational effectiveness, productivity, and throughput in a manufacturing operation with examples.

1.5 Differentiate between common operational availability calculation and Toyota's availability calculation, providing a simple case.

1.6 Using a case study, discuss how the role of direct labor cost impacts the total cost of a manufacturing system.

1.7 Discuss an application of Little's Law in a manufacturing system.

1.8 Review the necessity of WIP and its impact on a manufacturing operation.

1.9 Explain the differences between an optimization project and a continuous improvement project.

1.10 Analyze a production operation and identify its time elements (refer to Figure 1.8).

1.11 Compare the two definitions of cycle time with a real-world example and provide your insight into which one is more applicable.

1.12 Using an example from your workplace, discuss the type of process and associated KPIs and throughput concerns (refer to Table 1.1).

1.13 Review a manufacturing system from both a conversion view and composition view, providing an example.

1.14 Discuss system layout configurations for their characteristics, providing examples.

1.15 Comment on the opportunities and constraints of continuous throughput improvement on the production floor for distinct types of systems (refer to Figure 1.17).

1.16 Use an example to explain the characteristics of a batch process for multiple products.

1.17 Review situations of starvation and blockage in a production environment, providing examples.

1.18   Discuss the meaning of the standalone concept, providing an example.

1.19   Review considerations for reporting priority in operational states for a production line.

1.20   Discuss an application of the standalone concept to throughput indicators, such as operational availability and throughput rate, providing an example.

References

Aboutaleb, A., Kang, P.S., Hamzaoui, R. and Duffy, A. 2017. Standalone closed-form formula for the throughput rate of asynchronous normally distributed serial flow lines. Journal of Manufacturing Systems, 43, pp. 117–128.

Alithya 2023. 2023 Manufacturing sector trends: survey results and analysis. https://info.alithya.com/alithya-manufacturing-survey-2023. Accessed July 2023.

Clay, I. 2023. Recent U.S. manufacturing employment growth hides the sector's abysmal productivity performance, Information Technology and Innovation Foundation (ITIF). https://itif.org/publications/2023/02/23/recent-us-manufacturing-employment-growth-hides-the-sectors-abysmal-productivity-performance. Accessed June 2023.

Cunha, F., Dinis-Carvalho, J., and Sousa, R.M. 2023. Performance measurement systems in continuous improvement environments: obstacles to their effectiveness. Sustainability, 15(1), p. 867.

Fradkin, Y., Cordonnier, M., Henry, A. and Newton, D. 2017. Using an assembly sequencing application to react to a production constraint: a case study. SAE International Journal of Materials and Manufacturing, 10(3), pp. 316–319.

Hopp, W.J. and Simon, J.T. 1993. Estimating throughput in an unbalanced assembly-like flow system. The International Journal of Production Research, 31(4), pp. 851–868.

Koenigsberg, E. 1959. Production lines and internal storage — A review, Management Science, 5(4), pp. 410–433.

Kronos 2016. The future of manufacturing: 2020 and Beyond, IndustryWeek Special Research Report. https://www.nist.gov/system/files/documents/2016/11/16/iw_kronos_research_report_2016.pdf. Accessed April 2023.

Li, Y., Chang, Q., Brundage, M.P., Xiao, G. and Biller, S. 2013. Standalone throughput analysis on the wave propagation of disturbances in production sub-systems. Journal of Manufacturing Science and Engineering, 135(5), 051001.

Little, J.D.C. 1961. A proof of the queuing formula: $L = \lambda W$. Operations Research, 9(3), pp. 383–387.

Macmillan n.d. https://www.macmillandictionary.com/us/dictionary/american/throughput. Accessed April 2023.

Miller, T.D. 2021. Art of the Possible Handbook, Air Force Sustainment Left, United States Air Force. https://static.e-publishing.af.mil/production/1/af_sustainment_ctr/publication/afsch60-101/afsch60-101.pdf. Accessed September 2022.

Olaitan, O., Yu, Q., and Alfnes, E. 2017. Work in process control for a high product mix manufacturing system Procedia CIRP, 63, pp. 277–282.

Renna, P., Materi, S., and Ambrico, M. 2023. Review of responsiveness and sustainable concepts in cellular manufacturing systems. Applied Sciences, 13(2), p. 1125.

Sakai, H., and Li, J. 2020. New operational availability model to evaluate manufacturing throughput: advanced TPS for global production. Universal Journal of Management, 8(3), pp. 96–102. https://doi .org/10.13189/ujm.2020.080306.

Shafeek, H. and Marsudi, M. 2015. The relationship between batch size, throughput and utilization in manufacturing processes. International Journal of Engineering Trends and Technology, 28(3). https://doi.org/10.14445/22315381/IJETT-V28P221.

Simchi-Levi, D. and Trick, M.A. 2011. Introduction to "Little's law as viewed on its 50th anniversary". Operations Research, 59(3), pp. 535–535.

Skalli, D., Charkaoui, A., Cherrafi, A., Garza-Reyes, J.A., Antony, J. and Shokri, A. 2022. Industry 4.0 and Lean Six Sigma integration in manufacturing: a literature review, an integrated framework and proposed research perspectives. Quality Management Journal, 30, pp. 1–25.

Valdés, R. and Leoncio, J. 2003. Constraint analysis and throughput improvement at an automotive assembly plant. Master's Thesis, Department of Mechanical Engineering and the Sloan School of Management, Massachusetts Institute of Technology, 94 pages.

Yelles-Chaouche, A.R., Gurevsky, E., Brahimi, N. and Dolgui, A. 2021. Reconfigurable manufacturing systems from an optimisation perspective: a focused review of literature. International Journal of Production Research, 59(21), pp. 6400–6418.

2

System Performance Metrics

2.1 Performance Measurement

Evaluating and improving a manufacturing system's performance necessitates measuring its operational effectiveness and efficiency. Different aspects of manufacturing operations necessitate diverse performance measurements, including economic, resource, and environmental efficiencies [Castiglione et al. 2022].

Throughput measurements and indicators can vary across industries. For instance, in the chemical industry, the throughput in a reaction process is often measured by the heat transfer rate between the reactor contents and the jacket. The throughput is about the flow rate of heating or cooling media. In energy-intensive industries, energy efficiency should be considered when optimizing throughput performance [Alaouchiche et al. 2020]. A study on hand-tool manufacturing utilized a KPI that specifically focused on the total loss of time, process, and material [Shiau and Wang 2021].

This chapter delves into different indicators and methods for measuring manufacturing system throughput.

2.1.1 Direct Throughput Indicators

2.1.1.1 Production Count and Rate

The simplest way to measure a system's throughput is to count the number of products manufactured within a specific period. For example, a shift produced 560 vehicles. The count indicator is frequently utilized to assess system performance by comparing the actual output with a set production target. Figure 2.1 shows an example of the comparison between the actual count of produced items and the target production on the shop floor.

To ensure a valid comparison between actual and target outputs, exclude scheduled nonproduction times (such as breaks) from the calculations. It is important to note that product quality is normally not factored into the production count, even if some manufactured products do not meet specifications. Therefore, relying solely on the production count may not accurately reflect the true throughput performance of a manufacturing system.

As discussed in Chapter 1, system throughput often refers to the rate of product units built per unit of time. For instance, a throughput rate (TR) can be expressed as jobs per hour (JPH). A TR serves as an effective performance index for monitoring fluctuation and variation, thereby providing valuable information for root cause analysis.

Figure 2.2a shows an example of a throughput accumulation chart, and Figure 2.2b illustrates the deviation chart. In both charts, the dashed line represents the planned target throughput,

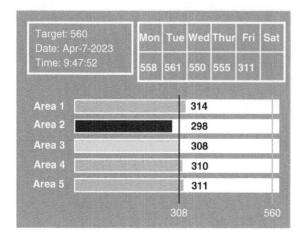

Target: 560 Date: Apr-7-2023 Time: 9:47:52	Mon	Tue	Wed	Thur	Fri	Sat
	558	561	550	555	311	

Area 1 314
Area 2 298
Area 3 308
Area 4 310
Area 5 311

308   560

Figure 2.1 System throughput count display board.

whereas the solid line denotes the actual performance. The flat steps in chart (a) of the dashed line indicate scheduled break times during production, which are required when people are involved in operations.

Accumulation chart Figure 2.2a displays the difference between the planned and actual amounts over time. In this case, the actual throughput lags the planned target throughput. Overtime work is then used over the last break time to compensate for and offset this loss.

On the contrary, in the deviation chart, Figure 2.2b, the slope of the actual throughput line suggests that the system's actual TR is lower than its design (required) TR. This chart can assist operational management in identifying instances when the system falls behind and needs change and improvement.

The aforementioned throughput indicators, such as quantity and TR, offer a broad overview of production output and are employed on the production floor. For an in-depth understanding of system performance and to pinpoint potential enhancements, it is essential to delve beyond basic indicators and scrutinize underlying causes and factors.

2.1.1.2 System Availability Metric

As outlined in Chapter 1, system operational availability (A) measures the operational percentage of a system during scheduled work time:

$$A = \frac{\text{Actual production time}}{\text{Planned production time}}.$$

Several variations in system operational availability exist. One variation is volume attainment (VA), based on production quantity:

$$VA = \frac{\text{Actual produced units}}{\text{Planned output units}}(\%).$$

VA directly measures the quantity of production output within a specific period. For instance, if a system was planned to produce 500 units and produced 475 units, then:

$$VA = \frac{\text{Actual produced units}}{\text{Planned output units}} = \frac{475}{500} = 95\%.$$

As VA is an overall indicator, it is recommended to use it in conjunction with other measures for a more comprehensive understanding of system performance.

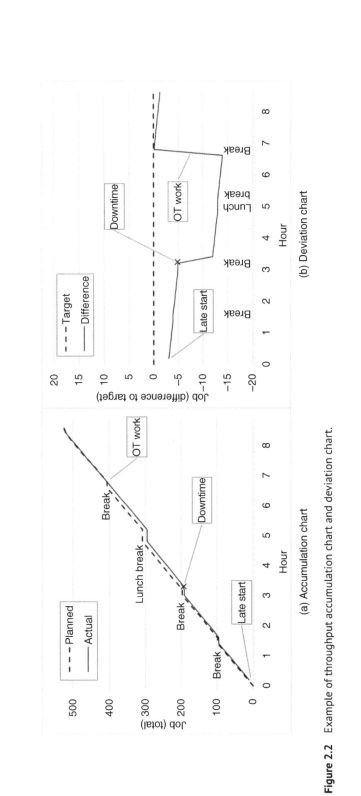

Figure 2.2 Example of throughput accumulation chart and deviation chart.

Table 2.1 Example of volume attainment vs. schedule attainment.

System	VA (by unit, %)	SA (by unit and time, %)	A
...			
5	95	98	0.970
6	103	105	0.981
7	100	95	1.053
...			

Production time plays a crucial role in throughput measurement, considering both the actual and the planned time. Schedule attainment (SA) is another indicator that considers the ratios of actual to planned production units and time. It can be calculated as:

$$SA = \frac{\dfrac{\text{Actual produced units}}{\text{Actual production time}}}{\dfrac{\text{Planned output units}}{\text{Planned production time}}} = \frac{\text{Actual produced units}}{\text{Planned output units}} \times \frac{\text{Planned production time}}{\text{Actual production time}}(\%).$$

Based on the definitions of A and VA, we can simplify the SA equation as follows:

$$SA = VA \times \frac{\text{Planned production time}}{\text{Actual production time}} = VA \times \frac{1}{A}.$$

This relationship between VA and SA underscores the role of A. In addition, since A is typically less than 1, SA will be greater than VA. SA would be equal to VA if A equals 1. Table 2.1 provides an example with three scenarios for further discussion.

- System 5 appears to be normal, aligning with the aforementioned relationship.
- System 6 has a VA and SA exceeding 100%, suggesting it operated at a faster rate.
- System 7 has a VA of 100%, indicating perfect performance. However, the value of A is questionable as it cannot exceed 1, suggesting that System 7 utilized overtime work to meet the scheduled production (VA = 100%) without updating the work time in the system.

Subsection Summary: A few direct throughput indicators highlighted, including production count, TR, VA, and SA, offer valuable insights into manufacturing system performance.

2.1.2 KPIs of Operations

2.1.2.1 Variety of KPIs

KPIs form the core of performance measurement and target setting of any business system. They can be applied to various aspects of a business, including finance, process effectiveness, marketing, and sales.

Various aspects of a manufacturing system can demonstrate the system's operational performance. Figure 2.3 illustrates six main areas or factors contributing to system throughput. Each area encompasses several dimensions or items. Table 2.2 provides a snapshot of KPI applications across six areas in the context of a repetitive manufacturing process.

For effective throughput management, each KPI should have a predetermined target value based on historical data, business mission, and departmental goals. The actual performance is then presented as a percentage of the target, with different color-coded levels to facilitate easy recognition

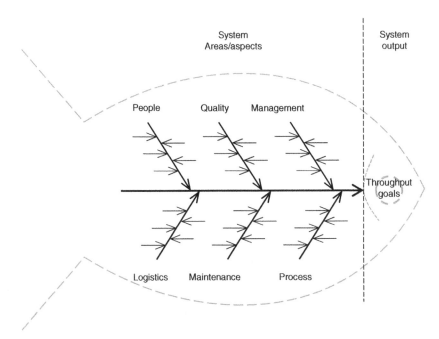

Figure 2.3 Main contributing areas/factors to throughput performance.

Table 2.2 Manufacturing KPIs in six different areas/aspects.

Area	Item	Target	%	Area	Item	Target	%
Management	Masterplanning			Logistics	Material flow plan		
Management	Production control			Logistics	On-time delivery		
Management	Employee safety			Logistics	Line side presentation		
Process	Line cycle time			Logistics	Fork truck/AGV utilization		
Process	Standardizedwork			Quality	First-timequality		
Process	Visual management			Quality	Inspections and audits		
Process	Supervisor call (Andon)			Quality	Process capability		
Process	Error/mistake proofing			Quality	Andon cord activated times		
Maintenance	Meantimebetweenfailure			People	Operatortraining		
Maintenance	Mean time to repair			People	Team rotation		
Maintenance	Unscheduled downtime			People	Problem-solving skills		
Maintenance	Spare part plan			People	Workplace organization (5S)		

and follow-up. For example, green for meeting or exceeding targets, yellow for near targets, and red for below targets. In addition, graphic charts can also be used to visualize the gaps between current operations and the best achievable values.

In addition to commonly used KPIs, a literature review revealed other types of performance measurements such as makespan (time taken to process a fixed number of jobs), profit rate, lead time, WIP, customer satisfaction, market share, arrival rate, mix proportion, and set up [Battesini et al. 2021]. The selection of KPIs is influenced by factors such as production volume, product variety, system size, and the level of automation. Further discussion on KPI selection is conducted in subsection 2.4.1.

2.1.2.2 Discussion of KPI Applications

When focusing on a specific production process or area, it is essential to select relevant KPIs from the system-level KPIs and customize them to fit local requirements. This approach establishes connections between specific processes or subsystems and the entire system. For example, a particular department may choose a few relevant system KPIs, such as quality rate and downtime, to showcase subsystem KPIs supporting the system and aligning with their operations and responsibilities.

Beyond routine monitoring, the gaps between current KPI values and targets can drive improvement tasks and projects, influencing planning and prioritization. Target values can be set at two levels: industry benchmarks and internal best records of a shop. For instance, a machining process might have a target MTTR (mean time to repair) of five minutes based on industry benchmarks and seven minutes as the internal best record.

The radar chart in Figure 2.4 serves as an example, illustrating a customized selection of KPIs. The chart compares the current performance status (marked with "◯") against the targets ("★"), highlighting gaps and suggesting improvement efforts to close them.

It is important to note that the significance of KPIs and their alignment with managerial goals, such as system throughput, can vary. While some KPIs are directly related to the system throughput, others may indirectly reflect some aspects of the system performance.

Figure 2.5 provides another example of a production line, highlighting four KPIs represented in the chart. Three of these KPIs — operational availability, speed performance, and product quality — are elements of overall equipment effectiveness (OEE). As a comprehensive metric, OEE is discussed in depth in Section 2.2.

Figure 2.4 Example of KPI radar chart of a small production area.

66.2 JPH	92.5%	95.1%	98.3%
Target	Target	Target	Target
Throughput rate	Availability	Performance	Quality

Figure 2.5 Example of KPI monitoring chart of a production line.

The frequency of KPI updates can vary based on their importance to manufacturing systems and the availability of data. While some KPIs may require daily updates for real-time monitoring with real-time monitoring systems, others may be updated weekly.

2.1.2.3 KPIs in Throughput Management

KPIs can directly impact system throughput. The first-time quality, a KPI, is directly linked to system throughput. Implementing an Andon (or Jidoka in the Toyota Production System) function, aimed at enhancing quality assurance, may occasionally slow down operations but contributes significantly to the overall throughput by minimizing rework and defects. Another example is the KPI of on-time delivery rate of inbound logistics, which can directly affect system throughput. A recent study shows that importing hybrid logistics infrastructures into existing material handling systems can reduce cycle time [Horn and Podgorski 2021].

From a managerial standpoint, it is important to convert throughput performance and its improvements into measurable outcomes, particularly contributions to financial gains. Figure 2.6 illustrates the relationships between three types of manufacturing KPIs: process KPIs, output KPIs, and outcome KPIs.

Process KPIs are specific to different areas within a manufacturing system. Analyzing process KPIs allows for a process-oriented perspective, helping identify root causes and improvement opportunities.

Output KPIs, such as TR and quality rate, are focused on results. While monitoring these is essential to understand the status, a balanced approach involves analyzing both result-driven (output KPIs) and process-oriented (process KPIs) perspectives.

The conversion of output KPIs, such as TR and quality rate, into outcome KPIs is important but has not yet been widely practiced. Subsection 2.4.3 has an in-depth discussion of the importance and challenges of the conversion.

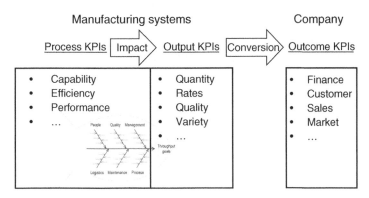

Figure 2.6 Relationships between the three types of manufacturing KPIs.

To achieve manufacturing excellence, a balanced approach that considers both result-driven and process-oriented perspectives is essential. Focusing on details, analyzing data, and understanding the intricacies of different KPIs contribute to effective throughput management by identifying root causes and opportunities to improve.

Subsection Summary: Various KPIs in six key aspects are outlined for assessing manufacturing performance. A selecting KPI process is explored from the system level to local processes, emphasizing how KPIs apply to throughput management.

2.1.3 Review of ISO 22400 KPIs

2.1.3.1 Common KPIs

The manufacturing industry encompasses numerous processes and practices, leading to a wide range of performance indicators. ISO 22400:2014 is a comprehensive standard for manufacturing operational management that provides valuable guidelines.

ISO 22400:2014, which was reviewed in 2023, aims to standardize data exchange protocols within the manufacturing industry. It offers a set of 34 KPIs for manufacturing operations management, covering their descriptions, formulas, units of measurement, ranges, trends, frequency, audience, and production method [ISO 2014]. This standard, based on ISO/IEC 62264, serves as a valuable baseline for performance measurement.

While ISO 22400 provides a standardized framework for KPIs, it may be better suited for repetitive manufacturing processes than for other types of processes. Studies have indicated that readers might encounter challenges in fully comprehending the standard, which can limit its practical applications in real-world scenarios [Varisco et al. 2018]. Therefore, there is room for improvement in terms of clarity, precision, and completeness across all manufacturing operations. When a new KPI is needed, ISO 22400 allows the proposal of new metrics with detailed specifications on timing, audience, formulas, units, usage, etc.

Apart from ISO 22400, several international and industrial standards define metrics for assessing process effectiveness and efficiency in quality and continuous improvement fields. These include the following:

- ISO 9001:2015 [ISO 2015]
- ISO 9004:2018 [ISO 2018]
- IATF 16949:2016 [IATF 2016]

In summary, while ISO 22400 and other standards serve as a foundational KPI framework, its applications may require adjustments to suit special manufacturing processes. The combination of multiple standards can insightfully guide the development of KPIs. As the manufacturing landscape continues to evolve, efforts should be made to continuously improve by refining and expanding the existing KPIs to enhance performance assessment and decision-making across the industry.

2.1.3.2 KPI Categorization

There are several main aspects of manufacturing system performance such as production, maintenance, inventory, and quality. The KPIs of ISO 22400 may be categorized into one of the four groups, as shown in Table 2.3 [Weiss et al. 2013]. These KPIs can be categorized in several ways, including grouping into areas such as overall effectiveness, capacity, quality, maintenance, and inventory.

Table 2.3 Categorization of common manufacturing KPIs [Adapted from Weiss et al. 2013].

KPI	Production	Maintenance	Inventory	Quality
Actual to planned scrap ratio				√
Allocation efficiency	√			
Allocation ratio	√			
Availability	√			
Comprehensive energy consumption	√			
Corrective maintenance ratio		√		
Critical machine capability index	√			
Critical process capability index	√			
Effectiveness	√			
Equipment load rate	√			
Fall off ratio				√
Finished goods ratio	√			
First pass yield				√
Integrated goods ratio	√			
Inventory turns			√	
Machine capability index	√			
Mean operation time between failures		√		
Mean time to failure		√		
Mean time to restoration		√		
Net equipment effectiveness index	√			
Other loss ratio			√	
Overall equipment effectiveness index	√			
Process capability index	√			
Production loss ratio	√			
Production process rate	√			
Quality ratio				√
Rework ratio				√
Scrap ratio				√
Set up rate	√			
Storage and transportation loss ratio			√	
Technical efficiency	√			
Throughput rate	√			
Utilization efficiency	√			
Worker efficiency	√			

Furthermore, a KPI may fall into more than one category. For instance, the "Availability" KPI is categorized under Production in the table. In addition, it can be applied to the Maintenance category as it can assess the effectiveness of maintenance in ensuring equipment uptime. Similarly, quality KPIs such as "Fall off ratio," "First pass yield," "Quality ratio," "Rework ratio," and "Scrap ratio" also significantly impact production operations.

The versatility of these KPIs enables manufacturers to effectively analyze and improve various aspects of their production, maintenance, inventory, and quality management. By considering the relevance of these KPIs to multiple categories, manufacturing professionals can develop a more comprehensive approach to performance evaluation and continuous improvement in their respective industries.

2.1.3.3 Throughput KPIs

Among the 34 KPIs in ISO 22400, some are specifically related to system throughput output, while others address various aspects of operational management such as effectiveness, indirect costs, and utilization efficiency. This book does not examine all KPIs in detail but instead focuses on the 12 most throughput-related KPIs along with the corresponding discussion in this book (Table 2.4).

These KPIs are thoroughly discussed throughout the book, with their primary or first-time discussion subsections cross-referenced in the right column of Table 2.4. While some other KPIs, such as set-up ratio and worker efficiency, are also related to throughput performance, they are not specifically addressed in this book.

Certain operations or industries may need additional throughput KPIs. For example, labor-intensive operations may focus on factors such as labor cost, its percentage of the total cost, and productivity. These additional KPIs can be developed, referring to the KPIs in ISO 22400 and the considerations addressed in this book.

Subsection Summary: KPI categorization and common throughput KPIs are comprehensively reviewed, referencing ISO 22400 and other standards. The 12 throughput KPIs serve as pivotal metrics for evaluating and enhancing system throughput.

Table 2.4 Throughput KPIs and discussion subsections.

	ISO22400 KPI	Most related subsection
1	Actual to planned scrap ratio	4.3.1 (quality to throughput)
2	Availability	1.1.2 (throughput performance)
3	Corrective maintenance ratio	5.3.2 (maintenance effectiveness)
4	First pass yield	4.3.2 (quality analysis)
5	Mean operation time between failures	5.1.2 (MTBF and MTTR)
6	Mean time to failure	5.1.2 (MTBF and MTTR)
7	Mean time to restoration	5.1.2 (MTBF and MTTR)
8	Overall equipment effectiveness index	2.2.2 (introduction to OEE)
9	Quality ratio	4.3.1 (quality to throughput)
10	Rework ratio	4.3.1 (quality to throughput)
11	Scrap ratio	4.3.1 (quality to throughput)
12	Throughput rate	1.1.2 (throughput performance)

2.2 Manufacturing System OEE

Most KPIs present individual factors, as discussed earlier. Composite KPIs integrate several individual factors into a single indicator. One commonly used composite KPI is OEE, which combines availability, performance, and quality metrics.

2.2.1 Three Major Metrics

2.2.1.1 Availability Element

As discussed in Chapter 1, operational availability (A) is an important and straightforward metric for gauging system throughput performance. Denoting actual operating time as t_a and planned production time as t_p, the commonly defined A is:

$$A = \frac{\text{Actual operating time}}{\text{Planned production time}} = \frac{t_a}{t_p}(\%),$$

while there are several calculations of operational availability, this book adheres to this specific definition.

2.2.1.2 Speed Element

Manufacturing operations may function at a speed slower than the design rate. In Toyota, slow operations are called "work delay time," which may have a greater impact on throughput than equipment downtime [Sakai and Li 2020].

Speed performance (P) is a metric gauging the speed or rate of a system's operation. It is computed as the ratio of actual units produced (u_a) to the planned units (u_p) during a system's uptime:

$$P(\text{unit based}) = \frac{\text{Actual units produced}}{\text{Planned units in uptime}} = \frac{u_a}{u_p}(\%).$$

For example, if operating at its design speed, a system is expected to produce 5200 units during a week. However, only 5000 units were produced. In this instance, the speed performance, P, can be calculated as $5000/5200 = 96.15\%$.

In certain manufacturing operations, such as repetitive or discrete-product processes, cycle time (CT) can be used instead of unit in the calculation.

$$P(\text{cycle time based}) = \frac{\text{Design CT}}{\text{Actual CT}} = \frac{CT_d}{CT_a}(\%).$$

The CT-based calculation can be favored because it does not involve unit counts, allowing for a quick evaluation. On the contrary, unit-based speed performance calculation may be chosen for its simplicity in conveying an overall evaluation, only when in the absence of downtime during a period. In-depth CT discussion is conducted in Section 6.3 of Chapter 6.

2.2.1.3 Quality Element

Assessing product quality is of utmost importance, as only good products contribute to the system's output. A measure of product quality is the quality rate, quantified as the ratio of good units produced (u_g) to the total produced units (u_a):

$$Q = \frac{\text{Good units produced}}{\text{Total units produced}} = \frac{u_g}{u_a}(\%).$$

For instance, a manufacturing system produced 5000 units in a week, with 4890 units meeting the quality requirements. In this case, the quality rate $Q = 4890/5000 = 97.8\%$. This is referred to as "first-pass yield" or "first-time quality," given that no repair and reprocessing are involved. After repairs and reprocessing, the resulting quality rate is termed the "overall quality rate." Chapter 4 delves deeper into the topics of quality and its impact on system throughput performance.

Subsection Summary: Three key metrics – operational availability, speed performance, and quality rate – are the main gauges for evaluating a manufacturing system's performance. Combining the metrics enables a holistic understanding of a system's overall performance. Further discussion will follow.

2.2.2 Introduction to OEE

2.2.2.1 Concept of OEE

Introduced in the 1980s by Nakajima [1988], OEE is a composite KPI, integrating three metrics – operational availability, speed performance, and quality rate – into a single measure. Originally developed for a single machine, the concept of OEE has been applied to various sizes of manufacturing systems.

This unified metric provides a comprehensive view of operational effectiveness, as the product of operational availability, speed performance, and quality rate.

$$\text{OEE} = A \times P \times Q\,(\%).$$

For instance, if the operational information of a shift operation is known (refer to Table 2.5), the corresponding OEE is:

$$\text{OEE} = A \times P \times Q = \frac{t_a \times \text{CT}_d \times u_g}{t_p \times \text{CT}_a \times u_a} = \frac{6.85 \times 45.6 \times 489}{7.5 \times 46.7 \times 500} = 87.22\%.$$

Being a KPI, OEE should have a target value or benchmark for comparison in evaluation and improvement consideration, which is discussed in subsection 2.2.3.3.

2.2.2.2 OEE Interpretation

The three OEE elements are often assumed to be independent. Under this assumption, the relationships between OEE and the elements are straightforward. Figure 2.7 shows the relationships between OEE and its individual elements for conceptual understanding. While, in practice, the independence assumption may not always hold, as will be discussed later.

Table 2.5 Example of a system's operational information.

OEE element	Parameter	Value
$A = \dfrac{t_a}{t_p}$	Actual operating time (t_a)	6.85 hours
	Planned production time (t_p)	7.5 hours
$P = \dfrac{\text{CT}_d}{\text{CT}_a}$	Design cycle time (CT_d)	45.6 seconds
	Actual cycle time (CT_a)	46.7 seconds
$Q = \dfrac{u_g}{u_a}$	Good units produced (u_g)	489 units
	Total produced units (u_a)	500 units

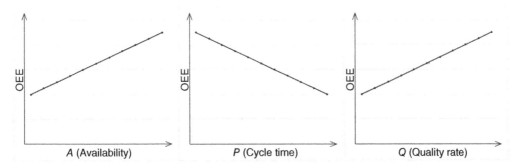

Figure 2.7   Relationships between OEE and its individual elements.

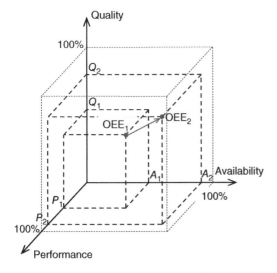

Figure 2.8   The 3D representation of OEE changes resulting from the three elements.

Furthermore, the three elements can concurrently change, collectively affecting OEE. In such cases, the relationship between OEE and its elements can be represented in a 3D graphic, as shown in Figure 2.8. The figure illustrates an OEE change resulting from concurrent changes in the three elements, transitioning from OEE_1 to OEE_2. While 3D OEE illustration is rarely done due to its complexity.

The OEE value includes the three losses that contribute to overall manufacturing performance, indicating potential aspects and room for improvement, as shown in Figure 2.9:

- Availability loss $= 1-A$
- Speed performance loss $= 1-P$
- Quality loss $= 1-Q$

OEE and its three-element losses can be illustrated over a time frame (see Figure 2.10). Downtime and speed losses can be quantified as segments of operational time, although speed loss is often less apparent. Quality loss in rate or unit can be translated into an equivalent production time duration. Figures 2.9 and 2.10 help in understating OEE.

Subsection Summary: OEE, a comprehensive KPI, integrates three key metrics (A, P, and Q) into a single measure, offering insights for identifying losses within a system and facilitating potential improvements.

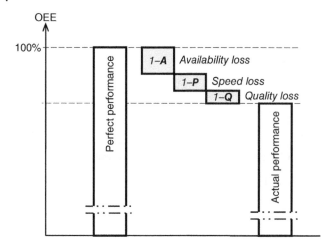

Figure 2.9 Contributions of three element losses to OEE.

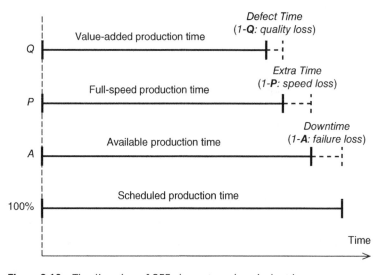

Figure 2.10 Timeline view of OEE elements and equivalent losses.

2.2.3 OEE Studies

2.2.3.1 OEE Accuracy Analysis

Since the losses of the three individual elements of OEE are $(1-A)$, $(1-P)$, and $(1-Q)$, their sum is $(1-A)+(1-P)+(1-Q)$. The corresponding system's performance, S_p, is given by:

$$S_p = 1 - [(1-A) + (1-P) + (1-Q)].$$

This is based on operational availability, performance speed, and product quality.

It is important to understand how OEE precisely represents a system's performance. A study shows that OEE values are remarkably close to S_p, when A, P, and Q are all above 90% [Tang 2019] as illustrated in Figure 2.11. OEE is remarkably close to and slightly overestimates a system's performance under different A, P, and Q. Therefore, OEE can serve as a good estimate of a manufacturing system's performance.

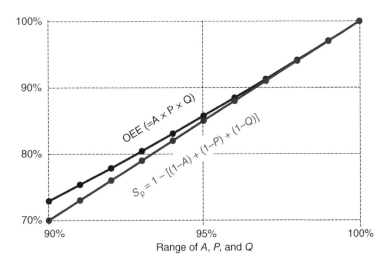

Figure 2.11 Comparison between system performance and OEE (Society of Manufacturing Engineers. 2013/with permission of Elsevier).

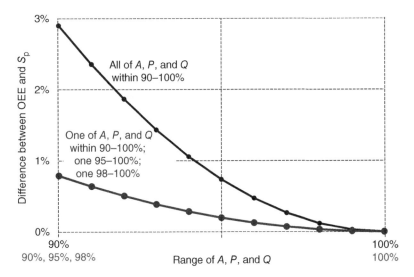

Figure 2.12 Estimation of difference between OEE and system performance (Society of Manufacturing Engineers. 2013/with permission of Elsevier).

When one or more elements are below 90%, OEE calculation can lead to an error of over 3% in estimating the system performance. Figure 2.12 presents two cases under different situations. It shows the difference in estimation when using OEE for S_p. Despite this, OEE can still serve as a good estimator of the system's performance.

For an overview, OEE serves as a valuable estimator for manufacturing system performance, closely reflecting the system's performance on availability, speed, and quality when A, P, and Q are above 90%.

2.2.3.2 OEE Change Estimation

In most cases, projects aimed at improving throughput result in small changes to A, P, and/or Q. In such instances, the resulting OEE changes can be quickly estimated without calculating OEE.

If the changes in operational availability (ΔA), speed performance (ΔP), and/or quality rate (ΔQ) are known, the corresponding OEE change (ΔOEE) can be expressed as:

$$\Delta OEE = OEE_2 - OEE_1 = (A + \Delta A) \times (P + \Delta P) \times (Q + \Delta Q) - A \times P \times Q.$$

Expanding this equation using the distributive property:

$$\Delta OEE = \Delta A \times P \times Q + A \times \Delta P \times Q + A \times P \times \Delta Q + \Delta A \times \Delta P \times Q + \Delta A \times P$$
$$\times \Delta Q + A \times \Delta P \times \Delta Q + \Delta A \times \Delta P \times \Delta Q.$$

Since the "Δ" in A, P, and/or Q from a project are typically small, usually less than 0.5%, the second-order and third-order terms of small changes on "Δ" in the equation can be negligible. Therefore, the OEE change can be approximated as:

$$\Delta OEE \approx \Delta A \times P \times Q + A \times \Delta P \times Q + A \times P \times \Delta Q.$$

For example, an improvement project increases Q by 0.4 percentage points, leading to corresponding improvements of $\Delta A = 0.1$ percentage points and $\Delta P = 0.05$ percentage points. The original values of A, P, and Q are 92%, 95%, and 98%, respectively. Based on the given info:

$$OEE_1 = A_1 \times P_1 \times Q_1 = 92\% \times 95\% \times 98\% = 85.65\%$$

$$OEE_2 = A_2 \times P_2 \times Q_2 = 92.1\% \times 95.05\% \times 98.4\% = 86.14\%$$

According to the estimation equation above,

$$\Delta OEE \approx 0.1 \times 0.95 \times 0.98 + 0.92 \times 0.05 \times 0.98 + 0.92 \times 0.95 \times 0.4 = 0.49\,(\%).$$

The estimated new OEE is 85.65% + 0.49% = 86.14%, which is the same as the calculated OEE_2 above.

Using this quick estimate is acceptable *if* the changes in all OEE elements are less than 0.5%, with the estimated error of OEE being less than 0.2%.

Furthermore, *if* A, P, and Q are already at good levels, for example, 95% or higher, the estimation can be simplified even further with an estimated error of less than 10%:

$$\Delta OEE \approx \Delta A + \Delta P + \Delta Q.$$

Using the same example:

$$\Delta OEE \approx \Delta A + \Delta P + \Delta Q = 0.1\% + 0.05\% + 0.4\% = 0.55\%.$$

For this case, the estimated relative error is approximately 12%, mainly due to A being only 92%. This simplified estimation is not recommended for this case. Therefore, estimating OEE can be convenient but conditional; all three elements should be better than 95% and individual changes should be less than 0.5%.

2.2.3.3 OEE Benchmarking

Being a KPI, OEE should have a target value for comparison, facilitating evaluation and improvement considerations. Notably, there is no standard target value or universal criteria for OEE, as it is contingent on the specific operation. OEE benchmarking serves as a crucial reference point. Table 2.6 provides good levels of OEE elements and overall OEE for large and complex manufacturing systems. Note that these values represent widely held beliefs based on common practices.

While these benchmarks offer valuable insights, the good levels of OEE can vary by industry. Table 2.7 provides examples of OEE levels in different manufacturing operations, offering specific references.

Table 2.6 Benchmarking reference of OEE elements and overall OEE values.

Element	Good level (%)
Operational availability	>92.0
Speed performance	>95.0
Quality rate	>97.0
OEE	>85.0

Table 2.7 Examples of manufacturing OEE levels.

Manufacturing operation	OEE level (%)	Reference
Ice cream production line	81	Tsarouhas [2020]
Blast furnaces in a year	73–88	Bhattacharjee et al. [2020]
Semiconductor components	76.5	Cheah et al. [2020]
Subassembly lines of car seats	73–79.5	Industrial ManagementDobra and Jósvai [2021]
15 press machines in a tool and die shop	75–77	Raju et al. [2022]
Faucet production facility in a year	88.9–91.2	Hatipoğlu and Akar [2022]

For large companies with multiple facilities, internal benchmarking serves as a reference for establishing OEE targets at different sites throughout the company.

In addition, comparing OEE year-over-year provides a genuine and meaningful measure of a system. When using OEE for subsystems, workstations, or individual equipment, it is important to note that different machines within a system can have varying impacts on overall performance. The presence of buffers inside a system can also influence the accuracy of OEE.

Subsection Summary: OEE can precisely indicate manufacturing system performance. Quickly estimating OEE changes becomes practical with small improvements in A, P, and Q, when they are at satisfactory levels. OEE benchmarking, both industry-wide and internal, provides valuable reference points for setting OEE targets.

2.3 Considerations on OEE

2.3.1 Weighting in OEE

2.3.1.1 Review of Existing Weighting

In the calculation of OEE, the three elements – A, P, and Q – are considered equally important. However, in certain situations, one element may carry more significance than others. Thus, it may be necessary to assign different weights to these elements based on their relative significance. Researchers have recognized the need to consider assigning varying weights to the three elements in OEE.

One approach to introducing weights is by using coefficients for each element in the OEE equation [Wudhikarn 2010]:

$$OEE_w = (w_A \times A) + (w_P \times P) + (w_Q \times Q),$$

where OEE_w denotes weighted OEE; w_A, w_P, and w_Q represent the weights for A, P, and Q, respectively. The author suggested determining these weights based on the ranking of the three elements, subject to the constraint:

$$w_A + w_P + w_Q = 1.$$

This approach is logical for weighting, although it can cause issues if equal weights are assigned to the three elements, or $w_A = w_P = w_Q = 0.33$. The corresponding OEE_w in this case becomes:

$$OEE_w = (0.33 \times A) + (0.33 \times P) + (0.33 \times Q) = 0.33 \times A \times P \times Q = 0.33 \times OEE.$$

This indicates that the weighted OEE is no longer comparable with the original (unweighted) OEE.

Another approach proposed by Raouf [1994] involves using the power (exponent) of the elements for weighting in the OEE equation:

$$OEE_w = (A^{w_A}) \times (P^{w_P}) \times (Q^{w_Q}).$$

Raouf suggested that the weights $w_A + w_P + w_Q$ should add up to 1. Like the previous weighted OEE approach, this weighting results in OEE_w not being comparable with the original OEE. When the weights of three elements are equal to 0.33, the resulting OEE_w becomes:

$$OEE_w = A^{0.33} \times P^{0.33} \times Q^{0.33} = (A \times P \times Q)^{0.33} = OEE^{0.33}.$$

From these two weighting approaches, assigning weights to the elements of OEE allows for considering their relative importance, but it leads to noncomparability with the original OEE calculation. Making weighted OEE comparable to unweighted OEE is important when both are used.

2.3.1.2 New Comparable Weighting

The weighted OEE elements should be assigned around one because OEE is the product of the three elements. A weight of one means no weight, the same as the original. Deviation from one indicates higher or lower importance. In this approach, the weights are assigned as the powers (exponents) of the elements:

$$OEE_w = (A^{w_A}) \times (P^{w_P}) \times (Q^{w_Q}).$$

Based on this principle, here is a new method to determine the weights for OEE elements with three steps [Tang 2023]:

1. Determine the ratings of each element on a scale of 1–10, with 10 being the most important. For example, assign ratings of 5, 2, and 9 to A, P, and Q, respectively.
2. Calculate a coefficient based on the ratings by dividing three by the sum of the three ratings. In this example, the sum is 16, so the coefficient is $\frac{3}{16}$.
3. Multiply the coefficient by each rating to obtain the corresponding weight. In this example, $w_A = 5 \times \frac{3}{16} = 0.9375$, $w_P = 2 \times \frac{3}{16} = 0.375$, and $w_Q = 9 \times \frac{3}{16} = 1.6875$.

By following these steps, the sum of three weights should be equal to 3. This weighting approach ensures that the weighted OEE maintains comparability with the original OEE. When the three elements are equally important, $w_A = w_P = w_Q = 1$, the weighted OEE is the same as the original OEE, which is the key advantage of this weighting method.

$$OEE_w = A^1 \times P^1 \times Q^1 = (A \times P \times Q)^1 = OEE$$

The weights should depend on the manufacturing system. Furthermore, these weight values can be subjective because they are based on the knowledge, judgment, and consensus of manufacturing professionals. Once determined, the weights should remain unchanged to ensure comparability with the previously established OEE values.

2.3.1.3 Interpretation of Weighted OEE

An example can be used to discuss the meaning of the new weighted OEE and compare it with conventional/unweighted OEE. Table 2.8 lists the A, P, and Q values, along with the calculated OEE and OEE_w for a manufacturing system over ten weeks.

During this period, three improvement projects were successfully implemented, focusing on operational availability in week 2, speed performance in week 5, and quality rate in week 7. For discussion purposes, the example used in the last subsection 2.3.1.2 is used here (the weights of 0.9375, 0.375, and 1.6875 for A, P, and Q, respectively).

Figure 2.13 illustrates the resulting OEE and OEE_w, which is a recommended practice for presenting unweighted OEE alongside weighted OEE.

Overall, with different weights on the elements, the resulting OEE_w differs from the original OEE. In this example, given that A is rated 5 out of 10, its improvement is similarly reflected in both OEE and OEEw, resulting in parallel changes. P is rated as 2 out of 10 (least important), its change is not significant in OEE_w. In contrast, Q is rated 9 out of 10 (most important), its change

Table 2.8 Discussion example of differences between OEE and OEE_w.

Week		1	2	3	4	5	6	7	8	9	10
Element	A (%)	89.0	88.8	91.4	90.9	91.0	90.8	91.2	91.5	91.3	91.4
	P (%)	92.0	92.2	92.1	92.5	92.3	95.1	95.3	94.8	94.9	95.2
	Q (%)	95.0	95.2	95.1	94.8	95.0	94.9	94.8	97.6	97.2	97.4
OEE	Original (%)	77.8	77.9	80.1	79.7	79.8	81.9	82.4	84.7	84.2	84.8
	Weighted (%)	79.7	79.9	81.9	81.2	81.5	82.1	82.3	86.6	85.8	86.3

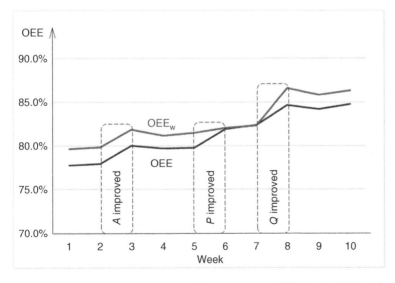

Figure 2.13 Example comparison of OEE and weighted OEE with individual element changes.

is amplified showing in OEE$_w$. Therefore, applying weights for OEE is to treat the three elements differently, reflecting their respective changing slopes in OEE$_w$.

Subsection Summary: There are weighting approaches for OEE but with comparable issues. A new comparable approach is introduced and discussed. The subjectivity of weight values and the consistency of applying weights need to be addressed in implementation.

2.3.2 Standalone OEE

2.3.2.1 Calculation of Standalone OEE

As explained, the standalone concept can reveal the true performance of a system by eliminating the external effects from the upstream and downstream systems. Standalone availability (A_{sa}) can be calculated using the following formula:

$$A_{sa} = \frac{\text{actual production time}}{\text{planned production time} - \text{starved time} - \text{blocked time}} = \frac{t_a}{t_p - t_s - t_b}.$$

Downstream operations can only block the current system's working. While upstream systems can influence the current operation's speed and quality. Therefore, without influences from the upstream systems, standalone speed performance (P_{sa}) and standalone quality rate (Q_{sa}) can be defined and calculated as follows:

$$P_{sa} = \frac{CT_d}{CT_a - CT_u},$$

$$Q_{sa} = \frac{u_g}{u_a - u_u},$$

where CT_u represents the CT loss due to upstream operations and u_u represents lost/defective units due to the upstream operations. For other variable denotations refer to subsection 2.2.1.

With A_{sa}, P_{sa}, and Q_{sa}, standalone OEE (OEE$_{sa}$) can be defined as:

$$OEE_{sa} = A_{sa} \times P_{sa} \times Q_{sa} = \frac{t_a \times CT_d \times u_g}{(t_p - t_s - t_b) \times (CT_a - CT_u) \times (u_a - u_u)}.$$

Table 2.9 lists the influences from upstream operations, with additional information to the parameters listed in Table 2.5, as an example for discussion.

The corresponding OEE$_{sa}$ can be calculated as:

$$OEE_{sa} = \frac{t_a \times CT_d \times u_g}{(t_p - t_s - t_b) \times (CT_a - CT_u) \times (u_a - u_u)}$$

$$= \frac{6.85 \times 45.6 \times 489}{(7.5 - 0.15 - 0.3) \times (46.7 - 0.3) \times (500 - 3)} = 93.95\%.$$

In this case, OEE$_{sa}$, significantly higher than the conventional OEE of 87.22% calculated in subsection 2.2.2.1, more accurately indicates the system's throughput performance. The difference, $93.95 - 87.22 = 6.37$ percentage points in this case, between standalone OEE and conventional OEE is the external influence, which is further discussed in the next subsection 2.3.2.2.

2.3.2.2 Implementation of Standalone OEE

Graphic illustration can better elucidate the difference between conventional OEE and OEE$_{sa}$. Figure 2.14 also indicates the three types of losses in the difference between the observed and actual throughput performance in terms of OEE.

Table 2.9 Example of a system's operational information with upstream influences.

OEE$_{sa}$ element	Parameter	Value
$A_{sa} = \dfrac{t_a}{t_p - t_s - t_b}$	Actual operating time (t_a)	6.85 hours
	planned production time (t_p)	7.5 hours
	Starved time (t_s)	0.15 hours
	Blocked time (t_b)	0.3 hours
$P_{sa} = \dfrac{CT_d}{CT_a - CT_u}$	Design cycle time (CT_d)	45.6 seconds
	Actual cycle time (CT_a)	46.7 seconds
	Loss due to upstream (CT_u)	0.3 seconds
$Q_{sa} = \dfrac{u_g}{u_a - u_u}$	Good units produced (u_g)	489 units
	Actual produced units (u_a)	500 units
	Defects due to upstream (u_u)	3 units

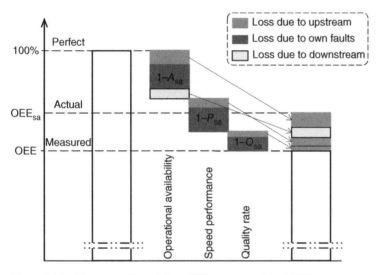

Figure 2.14 Elements of standalone OEE vs. conventional OEE.

By excluding the influences from neighboring systems, OEE$_{sa}$ provides valuable insights into actual system performance for managing and enhancing throughput. Figure 2.15 shows an example of OEE and OEE$_{sa}$ with the reasons for losses for three series-arranged assembly shops.

In this example, the OEEs and standalone OEEs do not align with each other. For instance, Body Paint has the lowest OEE, while Body Frame has the lowest OEE$_{sa}$. As OEE is a KPI for bottleneck identification in throughput improvement, the different observations on OEE and OEE$_{sa}$ raise a question, which will be studied in Chapter 3.

OEE$_{sa}$ sounds straightforward while analyzing the effects of upstream and downstream systems requires more information and effort. This task can be even more challenging for systems in parallel and hybrid configurations due to their complexity.

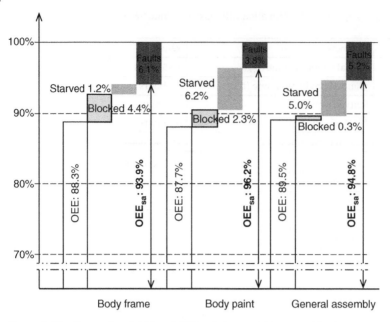

Figure 2.15 Conventional OEE and OEE$_{sa}$ of three vehicle assembly shops.

The implementation of OEE$_{sa}$ necessitates real-time recording of the status and effects of upstream and downstream systems. While calculating A_{sa} may not be challenging, obtaining immediate information and conducting analyses on how upstream systems influence P_{sa} and Q_{sa} is not always easy. In such cases, OEE$_{sa}$ can be simplified by excluding CT_u and u_u. The simplified OEE$_{sa}$ is:

$$OEE_{sa-simplified} = A_{sa} \times P \times Q = \frac{t_a \times CT_d \times u_g}{(t_p - t_s - t_b) \times CT_a \times u_a}.$$

For example, with the information provided in Table 2.9, the OEE$_{sa\text{-simplified}}$ is calculated to be 92.79%, compared with 93.95% for OEE$_{sa}$ and 87.22% for conventional OEE. Despite relying solely on A_{sa}, OEE$_{sa\text{-simplified}}$ still provides useful information for improving throughput. Concerns about speed performance and quality rate can be addressed when relevant information from upstream systems becomes available.

Subsection Summary: Standalone OEE can unlock in-depth throughput analysis to reveal a system's true performance by isolating external influences. Application examples highlight important considerations for the implementation of standalone OEE.

2.3.3 OEE Extensions

Applying the same logic used for OEE as a composite KPI, there are OEE variants tailored to different situations. Examples include overall throughput effectiveness (OTE) [Muthiah and Huang 2007] and overall plant effectiveness (OPE) [Muchiri and Pintelon 2008].

2.3.3.1 OEE for Labor Effectiveness – OLE

In manual operations and labor-intensive processes, the OEE approach can comprehensively measure workforce effectiveness or labor productivity. In such cases, OEE is also referred to as overall labor effectiveness (OLE), aimed at gauging the utilization, performance, and quality of the

Table 2.10 Comparison of element measurements of OEE and OLE.

Element	OEE	OLE
Availability (failure loss, %)	Operational availability, downtime, etc.	Workforce utilization, absenteeism, scheduling, indirect time, etc.
Performance (speed loss, %)	System speed, minor stoppage, etc.	Readiness, waiting, slow operation, etc.
Quality (quality loss, %)	Product quality rate, repair rate, scrap ratio, etc.	Operator error, setup error, maintenance error, etc.

workforce. The elements in OLE share similarities with those in OEE but with differing contents, as shown in Table 2.10.

For example, consider a manual process that started five minutes late for an eight-hour work shift. During this time, the process produced 190 parts against the target of 200, and five parts did not pass quality inspection. The calculation for OLE in this case would be as follows:

$$\text{OLE} = \frac{8 \times 60 - 5}{8 \times 60} \times \frac{190}{200} \times \frac{190 - 5}{190} = 91.54\%.$$

Since most manufacturing systems combine manual and automated operations, OEE and OLE can be used in tandem to evaluate distinct aspects of operational performance for throughput management. Figure 2.16 illustrates their applications at distinct levels of system automation. For instance, OLE may not be deemed necessary when a system is nearly fully automatic.

Applying OLE can face certain challenges related to scheduling, readiness, training, technology use, and other factors. Some of which may involve subjectivity. In addition, OLE values tend to vary more than OEE, given that OLE elements can be customized for a given process and work environment.

OLE can be used alongside other labor productivity KPIs, such as units per employee and hours per unit, introduced in Chapter 1. While these individual KPIs reflect certain aspects of workforce effectiveness, OLE provides a comprehensive overview.

OLE and other similar KPIs offer valuable insights into root causes, guiding throughput management to focus on addressing underlying issues when OLE is low. It is important to note that OLEs between operations at different facilities are often not fully comparable, as each facility may have significant variations in processes and work settings.

2.3.3.2 OEE with Time Utilization – TEEP

OEE is based on the scheduled work time; considering time utilization is another valid extension. For busy and continuous operations, total time – 24 hours a day – is a key factor in measuring

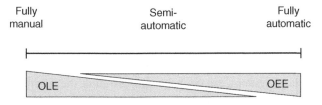

Figure 2.16 Applications of OEE and OLE at different levels of operation automation.

how effectively an operation uses time. In such cases, an additional element, time utilization (U), expressed as a ratio of the scheduled time over available time, is considered. For example, if a two-shift operation is planned a day, each consisting of 8.5 hours, then the time utilization of the operation for a five-day week is:

$$U = \frac{\text{Planned work time}}{\text{Total available time}} = \frac{2 \times 8.5 \times 5}{24 \times 5} = 70.83\%.$$

In total, five working days (or 24×5 hours) are used here as the available time. If seven days are considered instead, then the time utilization of the same operation is:

$$U = \frac{2 \times 8.5 \times 5}{24 \times 7} = \frac{85}{168} = 50.60\%.$$

Time utilization can be incorporated into OEE to form total effective equipment performance (TEEP) [Ivancic 1998], calculated as:

$$\text{TEEP} = A \times P \times Q \times U = \text{OEE} \times U.$$

TEEP is a performance metric for manufacturing operations, considering both system loss (measured by OEE) and time/schedule loss (measured by U). In TEEP, time utilization represents potential available time that could be utilized when needed. However, it is unrealistic for most manufacturing operations to achieve a U of 100%, unless they are running production around the clock without any breaks or downtime. Consequently, TEEP is typically lower than OEE under the same conditions.

For noncontinuous processes, such as repetitive and batch processes, TEEP yields a low percentage. From this perspective, it may be more appropriate to address time utilization as a separate factor, that is, using OEE and U at the same time, instead of using TEEP. For instance, the performance efficiency of an operation is 88.5% of OEE and 69.5% of U.

Subsection Summary: Beyond standard OEE, there are two important variants: OLE, assessing workforce effectiveness, and TEEP, incorporating time utilization into OEE. OEE variants offer specific, focused assessments for manufacturing systems.

2.4 Further Discussion of KPIs

2.4.1 KPI Selection

2.4.1.1 Overall Selection Process

Selecting appropriate KPIs is important for effectively measuring and managing manufacturing system performance. In this context, studies on KPI selection, such as an approach proposed for implementing and visualizing ISO 22400 KPIs [Ferrer et al. 2018], provide valuable insights.

The process of selecting KPIs is intricately tied to the unique characteristics of the manufacturing process and the specific improvement objectives a system aims to achieve. In their study, Lindberg et al. [2015] discovered that several industries faced challenges due to the absence of appropriate indicators for measuring and enhancing their performance. This emphasizes the critical role of well-chosen KPIs in facilitating meaningful performance assessments and targeted improvements.

The chosen KPIs should align with the objectives and provide meaningful insights into the system. Selecting suitable KPIs, especially for large systems, can be challenging. To guide the selection of suitable KPIs, especially for large systems, a recommended three-step approach is proposed here (see Figure 2.17).

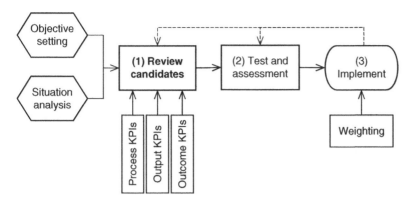

Figure 2.17 Process flow of KPI selection.

In the selection process, the first step is the core of the selection process. The second step involves testing, which includes the tryout and assessment of the newly selected KPIs to evaluate their effectiveness and suitability.

Consideration of weights for some KPIs, as discussed earlier using OEE as an example, can be optional and addressed after implementation. However, weighted KPIs may be difficult to comprehend and follow on the production floor.

The following subsections will delve into the selection process and related considerations in detail, providing valuable insights into the effective selection of KPIs for manufacturing systems.

2.4.1.2 KPI Candidate Review

In the selection process, the first step is to review the candidates of three types of KPIs (process, output, and outcome; refer to Figure 2.6). During this step, several questions must be addressed, including the following:

1. What are the suitably workable KPIs from the existing ones?
2. Are the candidate KPIs significant to the system?
3. What are the relationships among the KPIs?
4. How many KPIs are needed?

To initiate the process, the first question involves considering system performance and stakeholders based on existing KPIs. Many existing KPIs are listed in Appendix A, and if none fit the specific system, a new KPI can be developed.

The second question is to evaluate the significance of KPIs to the business and rank them. This, along with the third question, helps in better understanding the KPI candidates. Then, when considering the number of KPIs, it is preferable to have one to three KPIs for each objective to reduce potential conflicts arising from implying different improvement directions. All these considerations and results are illustrated in a map diagram, as exemplified in Figure 2.18.

This exemplary map of objective–KPI–data visualizes the relationships between the objectives, candidates, relationships, and requirements of KPIs. The left column of the map shows KPIs objectives. Then, the connections between objectives and KPIs can be built. The right column shows data sources for the KPIs.

It is important to note that this figure represents a conclusion of the KPI selection process, involving multiple iterations of brainstorming, analysis, and review. In addition, this example shows four objectives, which are typical ones. For additional objectives, more KPIs are needed.

Objectives (Significance)	KPI candidates	Descriptions (definition, area, etc.)	Data sources

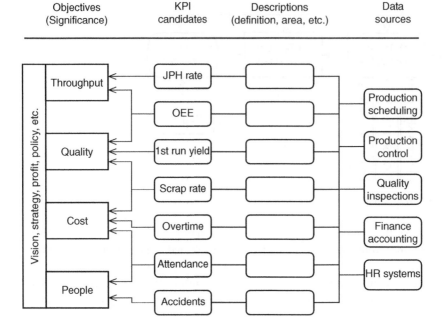

Figure 2.18 Objective–KPI–data relationship map for KPI selection.

2.4.1.3 Additional Selection Considerations

ISO 22400 lists 20 criteria for a KPI review to ensure KPI's usefulness in achieving various goals in a manufacturing operation [ISO 2014]. An article studied 14 KPI criteria [Hester et al. 2017], including quantifiable, relevant, predictive, standardized, verified, accurate, timely, traceable, independent, actionable, bought-in, understandable, documented, and inexpensive. These criteria or similar ones, particularly some important ones for a specific application, should be addressed when selecting or developing KPIs.

KPI candidates should be ranked based on their criteria. An evaluation form, exemplified in Table 2.11, serves as an example. In the table, each KPI is rated on the criteria on a scale of five, with five being the best. The total score is the sum of all individual ratings of a KPI candidate.

Documentation of KPIs, including their definitions, measurement process descriptions, and analysis guidance, should be completed before adaptation and implementation. The documentation is essential and provides a valuable reference for further modification and improvement.

Table 2.11 Example of KPI candidate comparison and selection.

Criteria	KPI-1	KPI-2	KPI-3	KPI-4	···
Quantifiable	5	4	4	3	
Significance to goals	4	5	3	5	
Accuracy	3	5	4	3	
Efficient reporting	4	4	3	4	
···					
Total score	16	18	14	15	

Once KPIs are evaluated and selected, they are ready for implementation in a real production environment.

Note that some KPIs have a certain level of abstraction and may have different calculations in companies with the same names. Consequently, it is important to understand the differences among practices of different companies when comparing the same KPIs and conducting a benchmarking study.

Subsection Summary: Selecting suitable KPIs is a comprehensive process, involving reviewing candidates, mapping objectives, ranking based on key criteria, validation testing, and documentation. Properly selected KPIs provide meaningful insights into the measurement and management of manufacturing system performance.

2.4.2 Considerations in KPIs

2.4.2.1 Relations among KPIs

System throughput is influenced by multiple factors or aspects, in confounding ways. Similarly, throughput KPIs can be impacted by other KPIs. Figure 2.19 provides an example of how these aspects and KPIs are interconnected, forming a network. In the figure, double lines denote strong influences between certain factors and their corresponding KPIs, impacting both throughput and each other. The main relationships are elaborated in the book's chapters and sections.

For instance, Kang et al. [2016] proposed a hierarchical structure to categorize and analyze KPI relationships. Luozzo et al. [2020] introduced a framework for identifying factors influencing ISO 22400 standard adoption. Zhu et al. [2018] conducted a gap analysis between industry needs and ISO 22400. These studies emphasize the importance of understanding KPI relationships within a system.

Certain KPIs relationships are inherently determined by system design and process planning. For these KPIs, there can be limited room for improvement in production management on the shop floor. Chapters 7 and 8 delve into crucial aspects of system design for throughput capacity.

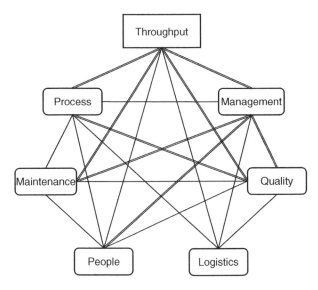

Figure 2.19 Interconnections among manufacturing aspects and related KPIs.

While some KPIs are used for routine reports, others are instrumental in analysis and problem-solving. In addition, integrating technical KPIs with critical management aspects such as finance, safety, health, and energy consumption is of significant interest and relevance. Ongoing research in this field is vibrant and actively pursued [Mohammed 2020].

Additional aspects of manufacturing performance, such as agile performance, are noteworthy. The original definition of agile manufacturing, as put forth by Kidd [1995] and Gunneson [1997], characterizes it as the ability to thrive and prosper in a competitive environment marked by continuous and unanticipated change. It involves the system's capability to respond rapidly to evolving markets driven by customer-centric product valuing. Subsequently, various definitions of agile manufacturing have emerged.

A key takeaway is that KPIs serve performance reporting purposes and improvement considerations. Understanding the interconnections among KPIs is crucial for effective KPI usage, as discussed in the following subsections.

2.4.2.2 Relation between Subsystem and System KPIs

A manufacturing system can be viewed as a complex hierarchical structure composed of multiple subsystems, where each subsystem contains several sub-subsystems and/or workstations that extend throughout the system. KPIs, such as OEE and TR, can be relevant and applied to all levels of an intricate system. However, a challenging question remains – how to comprehend the relationships among the same or similar KPIs across the various layers of a system?

Figure 2.20 provides an example of the dimensional relationship between the subsystem (producing door assembly) and the entire system (producing vehicle body assembly) [Tang 2022]. The dimensional quality of the door assemblies plays a role in influencing the quality, such as door gap and flushness, of the completed vehicle bodies.

Quality problems in a subsystem, like door assemblies in this example, may be exacerbated or partially mitigated within the manufacturing processes. Moreover, neighboring components, such

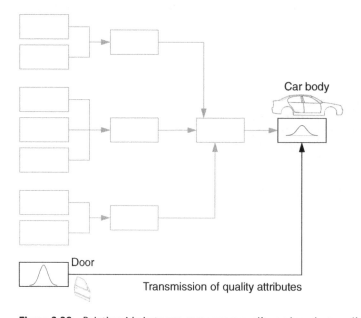

Figure 2.20 Relationship between component quality and product quality.

as the door opening on a vehicle body, may compensate to some degree for the gap issues from door dimensional quality. This logic can apply to other types of KPIs. For example, the OEE of the door assembly subsystem, a part of the vehicle body assembly system, may affect the OEE of the entire system.

Quality attributes and the corresponding variation transmissibility throughout a system is a common study topic in the manufacturing industry. A typical example is tolerance stacking analysis, widely used for assessing individual components right up to final product assembly. However, there is a need to address the transmissibility of KPIs. A subsystem's KPI may significantly impact the KPIs of the entire system or have no influence at all. Further discussion follows in the next subsection 2.4.2.3.

2.4.2.3 Subsystem–System KPI Relation Analysis

The relationship of a KPI between a subsystem and the entire system can be modeled as a reflection. This relationship can be estimated using a linear model, as illustrated in Figure 2.21 [Tang 2022].

With available datasets of a door assembly and final vehicle assembly, the KPI's relationship can be modeled as a reflection, which may be linear. A smaller slope implies a weaker relationship. This implies that the door assembly has less influence on the car door gap in this case. Furthermore, the reflection line can be nonlinear, shown as a dashed curve in the figure.

Discussions concerning other types of KPIs, such as OEE, including their analysis of relationships can adopt this approach. This relationship knowledge is highly valuable for enhancing the understanding of subsystem throughput capacities and for conducting bottleneck analyses within the system.

However, a challenge arises due to the absence of a theoretical model capable of accommodating these relationships, primarily due to the complexity of system configurations and variables involved. The nature of these relationships tends to be highly dependent on the specific case. To address this challenge, computer simulation becomes a valuable tool for analyzing these relations, as elaborated upon in Chapter 8.

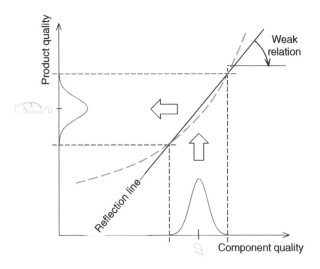

Figure 2.21 Linear reflection model of quality KPI transmissibility.

Subsection Summary: The relationships between KPIs from subsystems to the entire system are significant to manage system performance. Modeling such complex relationships can be challenging; a linear reflection model introduced can be used for such analysis.

2.4.3 Financial Implications of KPIs

2.4.3.1 Relation between KPIs and Finance

Financial health excellence is the primary goal of all businesses. Manufacturing systems, indicated by various KPIs, contribute to business goals. Figure 2.22 illustrates the link between manufacturing systems, KPIs, and financial profits. Profit should be a driving factor for KPI selection, as discussed earlier in this chapter.

Almost all throughput KPIs have financial implications. For instance, a drop in the TR can significantly impact profits. In the late 1980s, a General Motors (GM) plant operated at 57 JPH instead of the target 63 JPH, resulting in a loss of six vehicles per hour [Kohls 2020]. For a week of five days with two shifts per day and assuming a vehicle profit of $3000, the weekly loss would be $3000 × 6 (JPH) × 15 (hours per day) × 5 (days) = $1,350,000. Smaller products and components with small profit margins can also yield significant financial implications in volume production.

A mindset gap often exists between senior management and manufacturing practitioners. Management tends to use business language lefted around money and time while manufacturing practitioners primarily communicate using technical terms. To bridge this gap, manufacturing practitioners should present both technical successes and financial gains resulting from throughput improvements. Management can make informed decisions quickly based on throughput performance on the production floor with financial analysis.

Explicit financial and asset analyses of throughput are advisable to enhance this connection. Some financial terms based on throughput performance include:

- Operating profit before tax and interest (net profit) = Throughput – Operating expense
- Return on investment (ROI) $= \dfrac{\text{Net profit}}{\text{Inventory}}$
- Overall productivity $= \dfrac{\text{Throughput}}{\text{Operating expense}}$
- Inventory turns $= \dfrac{\text{Throughput}}{\text{Inventory}}$

Economic analysis is discussed further in the following subsections and Chapters 4 and 6.

2.4.3.2 Throughput Accounting

Accounting offers a crucial perspective on throughput management, aligning manufacturing performance with business objectives. In the context of throughput, accounting exercises underscore resource utilization, delivery efficiency, and profit analysis. Throughput-oriented financial

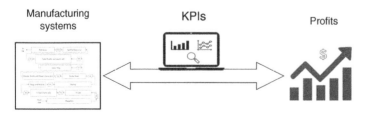

Figure 2.22 Manufacturing systems, operational KPIs, and finance/profit.

Table 2.12 Comparison between traditional accounting and throughput accounting.

Feature	Traditional accounting	Throughput accounting
Focus	Cost optimization	Maximizing throughput
Cost allocation	To products	Variable costs to products; fixed costs to operating expenses
Profitability measures	Net income and return on investment (ROI)	Throughput and return on investment
Decision-making	Driven by cost-cutting	Based on the impact on throughput
Overall perspective	Rule-based; adherence to standard practices	Flexible; focus on value and profit potential

analysis, which differs from traditional accounting practices, is effective in maximizing system throughput and business profit, as briefly summarized in Table 2.12.

For instance, traditional accounting lefts on product costs. While throughput accounting translates throughput-related concerns, such as resources, bottlenecks, improvement projects, and investment directions, into monetary terms. To calculate throughput, one can measure the sales revenue minus total variable costs and determine a ratio of return per factory hour [Kaplan Publishing 2017].

In the throughput accounting exercise, a unit cost model based on time bottleneck was proposed and explored [Myrelid and Olhager 2019]. Financial analyses can be performed on known constraints to measure and compare their impacts. The output of throughput accounting guides management decision-making based on the measurements of overall system capacity, performance, and the improvements of material flow through the entire system [Hilmola and Gupta 2015]. A case study found that traditional methods underestimated factors, such as demand variation and inventory, influencing decision-making [da Silva Stefano et al. 2022].

In practice, throughput accounting can complement traditional cost accounting rather than replace it. For example, Bragg [2007] proposed several throughput financial analysis scenarios from an accounting executive's perspective. Manufacturing professionals can diligently work with the accounting department and conduct a throughput-related financial analysis to support.

2.4.3.3 Justification for Throughput
With throughput accounting in mind, any actions and projects aimed at improving throughput undergo evaluation for their value and financial justification. This mythological thinking process can be illustrated using a case study.

In manufacturing, overtime work is used for one of two purposes: (1) to increase production output or (2) or make up for throughput loss. Conducting a financial analysis of overtime work under various scenarios helps better understand its financial implications.

Figure 2.23 illustrates a comparison of these two purposes as an example. When aiming to increase production output, the unit cost during an overtime work period is slightly higher due to a higher overtime work compensation. However, if there is production loss due to throughput or other types of issues, using overtime can offset the loss. In such cases, the unit cost is much higher, as the total produced units remain the same. This analysis assists in determining whether the overtime work is financially justified.

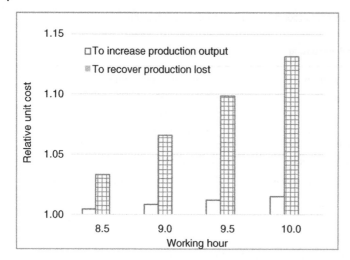

Figure 2.23 Cost implication of production overtime work.

A financial analysis of throughput gaps can aid in justifying the investment associated with a throughput improvement project. For instance, one might investigate how many weeks of over-time costs it would take to offset an additional investment in throughput improvement initiatives. Answers to such questions support decision-making in throughput management.

Several examples of ROI from throughput projects. GM managed to reduce overtime work while increasing sales of high-demand vehicles by throughput improvement, resulting in savings and revenue exceeding $2.1 billion [Alden et al. 2006]. Another success story comes from Trane U.S. Inc. It improved its throughput by 13% and reduced CT by 50%, leading to annual savings of over $0.7 million [Jensen et al. 2013]. These examples underscore the pivotal role of financial analysis in making investments in throughput improvement projects and demonstrate the benefits of throughput improvement.

Subsection Summary: Analyzing the financial implications of throughput changes is integral to throughput management. Throughput accounting and financial analysis can justify throughput improvement projects, as supported by real-world examples.

Chapter Summary

2.1. Performance Measurement

1. System throughput performance can be measured using various metrics against targets such as product quantity, output rate, attainment percentages, and operational availability.
2. Volume attainment is determined by the quantity of products produced, while SA considers both the quantity of products and the production time.
3. KPIs assess various aspects of system performance, including process effectiveness, quality, logistics, maintenance, and finance.
4. A predetermined target value should be established for each KPI for evaluation purposes.
5. ISO 22400 specifies 34 general KPIs for managing manufacturing performance, covering areas such as production, maintenance, quality, and inventory. Twelve of these KPIs are closely related to operational throughput performance.

2.2. Manufacturing System OEE

6. Operational availability, speed performance, and quality rate are crucial aspects of system performance.
7. Overall Equipment Effectiveness (OEE) is a comprehensive KPI that amalgamates three following KPIs: availability, speed, and quality.
8. OEE provides an accurate estimation of system performance when operational availability, speed performance, and quality rate are all above 90%.
9. Changes in OEE can be swiftly estimated from minor improvements in the three elements.
10. An OEE of 85% or more is deemed excellent in manufacturing. Benchmarking OEE against internal historical data or other similar operations aids in setting improvement targets.

2.3. Considerations on OEE

11. Weighting OEE elements allows for consideration of their relative importance but may result in incomparability with the original OEE.
12. A novel weighting approach is introduced to maintain comparability with the original OEE.
13. Standalone OEE unveils the system's true throughput performance by eliminating the influences from neighboring systems and external factors.
14. Overall Labor Effectiveness (OLE) is a variant of OEE that measures workforce utilization, performance, and quality in labor-intensive operations.
15. Total Effective Equipment Performance (TEEP) incorporates time utilization into OEE for continuous operations.

2.4. Further Discussion of KPIs

16. KPIs selection involves reviewing candidates, mapping objectives, ranking based on criteria, testing, and documentation. A four-step process is suggested.
17. Throughput KPIs are interconnected with other aspects of manufacturing performance such as process and quality, and their respective KPIs.
18. Subsystem and system KPIs have complex relationships and can be analyzed through linear reflection modeling and studied using computer simulation.
19. Throughput accounting examines throughput costs, investments, and bottlenecks in financial terms, which are explicitly linked to the business's bottom lines.
20. Financial analysis, including throughput achievements and losses, bottleneck effects, and root causes, justifies investments in improvement projects and helps convey impacts to management.

Exercise and Review

Exercises

2.1 A shop had planned to produce 450 units during a standard eight-hour shift. However, it ended up producing 451 units in 8.5 hours. Calculate the shop's volume attainment (VA) and schedule attainment (SA) performance. Comment on the results.

2.2 A plant operates two shifts, six days a week. Each shift is 8.5 hours long, inclusive of a half-hour lunch break and three 15-minute breaks for nonproduction activities. The plant is designed to have a throughput rate of 60 units per hour. Over the week, the plant manufactured 4440 units, out of which 90 required repairs. Also, there were events leading to 12 hours of unplanned downtime. Calculate the plant's speed performance of the week.

2.3 A production line is designed to have a cycle time of 60 seconds, but it operates at 61.2 seconds. Estimate the throughput rate of this production line. If the production runs for 7.5 hours per shift, determine the number of units lost a shift due to the speed performance rate.

2.4 Using the data provided in Exercise 2.2, calculate the OEE for the week.

2.5 A production system operated for 80 hours across five working days, with an OEE of 85%. Calculate its TEEP.

2.6 A quality improvement project can enhance the quality rate from 98.0% to 98.5%. It can also lead to a 0.1% improvement (92.0–92.1%) in operational availability and a 0.2% improvement (94.3–94.5%) in speed performance. Compare the estimated OEE and calculated OEE (refer to subsection 2.2.3).

2.7 For an operation, the significance of the three elements of OEE are assigned values of 7, 4, and 5, respectively. Over a week, these three elements were measured to be 91%, 99%, and 95%, respectively. Calculate the original OEE and the weighted OEE using the method introduced in subsection 2.3.1.2.

2.8 Continuing from Exercise 4, assume a standalone availability of 96%. What would be the simplified standalone OEE? Compare this with the results obtained in Exercise 2.2.

2.9 If a system's throughput is 0.5 JPH less than its target, and the unit profit is $1400, calculate the monetary loss for one week of production, assuming 75 working hours.

Review Questions

(The chapter covers these topics. For further discussion, it is recommended to seek additional information and examples. Diverse perspectives are encouraged.)

2.1 Discuss the differences between using an accumulation chart and a deviation chart to monitor system throughput (refer to Figure 2.2).

2.2 Review the applications of volume attainment (VA) and schedule attainment (SA) and provide an example.

2.3 Recommend one KPI from each of the six areas listed in Table 2.2 that would be most relevant to a specific manufacturing process of your choice and explain.

2.4 ISO 22400 recommends KPIs in four areas (refer to Table 2.3). Propose three KPIs from the production category that could effectively monitor throughput performance.

2.5 Select the top four or five KPIs from the 12 throughput KPIs listed in Table 2.4 for a particular production process and explain your reasoning.

2.6 Compare the advantages and disadvantages of using product quantity versus cycle time value for calculating the speed performance (P) metric in OEE and provide examples.

2.7 Should the repair and reprocessing of defective products be included in the quality rate (Q) of OEE? Please provide a rationale for your answer.

2.8 If the three elements of OEE change simultaneously but at different rates, will the OEE change linearly? Explain your reasoning.

2.9 What is the industry benchmark or best level for OEE in the industry you have selected?

2.10 In a specific scenario, which of the following losses would be the easiest to improve to enhance OEE: quality loss, speed loss, or failure loss? Provide an example and explanation.

2.11 Explain when it is necessary to incorporate weights into the three elements of OEE.

2.12 Discuss the accuracy of ΔOEE when its three elements undergo changes.

2.13 Evaluate whether standalone OEE is a practical and feasible metric to implement for a specific manufacturing process. Support your position with examples.

2.14 Compare OEE and OLE and provide an example along with an explanation.

2.15 Identify and examine an application of TEEP.

2.16 Discuss the process of selecting KPIs and considerations for a manufacturing system and provide an example.

2.17 Provide an example of a manufacturing system and describe how throughput, quality, and cost KPIs may interact with or influence each other within that process.

2.18 Assess the relationship between a subsystem KPI and the overall system KPI and suggest methods to study their relationship.

2.19 Discuss the application of throughput accounting principles in a continuous improvement project.

2.20 Evaluate the effectiveness of using overtime work to recover throughput losses. Discuss the benefits and drawbacks in terms of the total cost.

References

Alaouchiche, Y., Ouazene Y. and Yalaoui, F. 2020. Economic and energetic performance evaluation of unreliable production lines: An integrated analytical approach. IEEE Access, 8, pp. 185330–185345, https://doi.org/10.1109/ACCESS.2020.3029761.

Alden, J.M., Burns, L.D., Costy, T., Hutton, R.D., Jackson, C.A., Kim, D.S., Kohls, K.A., Owen, J.H., Turnquist, M.A., and Veen, D.J.V. 2006. General motors increases its production throughput. Interfaces (Providence), 36(1), pp. 6–25. https://doi.org/10.1287/inte.1050.0181

Battesini, M., ten Caten, C.S. and de Jesus Pacheco, D.A. 2021. Key factors for operational performance in manufacturing systems: Conceptual model, systematic literature review and implications. Journal of Manufacturing Systems, 60, pp. 265–282.

Bhattacharjee, A. Roy, S. Kundu, S. et al. 2020. An analytical approach to measure OEE for blast furnaces. Ironmaking and Steelmaking, 47(5), pp. 540–544. https://doi.org/10.1080/03019233.2018.1554348

Bragg, S.M. 2007. Throughput accounting: A guide to constraint management, ISBN 978-0471251095, Wiley.

Castiglione, C., Pastore, E. and Alfieri, A. 2022. Technical, economic, and environmental performance assessment of manufacturing systems: The multi-layer enterprise input-output formalization method. Production Planning & Control, 24 pp. 1–18.

Cheah, C.K., Prakash, J., and Ong, K.S. 2020. An integrated OEE framework for structured productivity improvement in a semiconductor manufacturing facility. International Journal of Productivity and Performance Management, 69(5), pp. 1081–1105. https://doi.org/10.1108/IJPPM-04-2019-0176

da Silva Stefano, G., dos Santos Antunes, T., Lacerda, D.P., Morandi, M.I.W.M. and Piran, F.S. 2022. The impacts of inventory on transfer pricing and net income: Differences between traditional accounting and throughput accounting. The British Accounting Review, 54(2), p. 101001.

Dobra, P. and Jósvai, J. 2021. OEE measurement at the automotive semi-automatic assembly lines. Acta Technica Jaurinensis, 14(1), pp. 24–35. https://doi.org/10.14513/actatechjaur.00576

Ferrer, B. R., Muhammad, U., Mohammed, W.M. and Martínez Lastra, J.L. 2018. Implementing and visualizing ISO 22400 key performance indicators for monitoring discrete manufacturing systems. Machines, 6(3), p. 39.

Gunneson, A.O. 1997. Transitioning to Agility: Creating the 21st Century Enterprise, Addison-Wesley Publications.

Hatipoğlu, S. and Akar, C. 2022. A new scoring approach to calculate overall equipment efficiency: A case study. Verimlilik Dergisi, 3, pp. 499–510. https://doi.org/10.51551/verimlilik.1055354

Hester, P., Ezell, B., Collins, A., Horst, J., and Lawsure, K. 2017. A method for key performance indicator assessment in manufacturing organizations. International Journal of Operations Research (Online), 14(4), pp. 157–167.

Hilmola, O.P. and Gupta, M. 2015. Throughput accounting and performance of a manufacturing company under stochastic demand and scrap rates. Expert Systems with Applications, 42(22), pp. 8423–8431.

Horn, G.W. and Podgorski, W. 2021. Throughput improvements via logistics in current semiconductor factories In 2021 China Semiconductor Technology International Conference (CSTIC), pp. 1–2. https://doi.org/10.1109/CSTIC52283.2021.9461453.

IATF. 2016. IATF 16949:2016 Quality Management Systems standard for the Automotive industry. International Automotive Task Force (IATF). https://www.iatfglobaloversight.org/.

ISO. 2014. ISO 22400-1:2014 automation systems and integration—key performance indicators (KPIs) for manufacturing operations management—Part 1: Overview, concepts and terminology, Part 2: Definitions and descriptions. Geneva, Switzerland.

ISO. 2015. ISO 9001:2015 Quality management systems. Geneva, Switzerland.

ISO. 2018. ISO 9004:2018 Quality management. Geneva, Switzerland.

Ivancic, I. 1998. Development of Maintenance in Modern Production, Euromaintenance '98 Conference Proceedings, Dubrovnik, Hrvatska.

Jensen, J.B., Ahire, S.L., and Malhotra, M.K. 2013. Trane/Ingersoll Rand combines lean and operations research tools to redesign feeder manufacturing operations. Interfaces, 43(4), pp. 325–340.

Kang, N., Zhao, C., Li, J., and Horst, J.A. 2016. A hierarchical structure of key performance indicators for operation management and continuous improvement in production systems, International Journal of Production Research, 54(21), pp. 6333–6350. https://doi.org/10.1080/00207543.2015 .1136082

Kaplan Publishing. 2017. F5 Performance Management, ISBN: 9781784158101, Kaplan Publishing (United Kingdom).

Kidd, P.T. 1995. Agile Manufacturing: Forging New Frontiers, Addison-Wesley Longman Publishing Co., Inc.

Kohls, K. 2020. GM's Throughput Improvement Process. https://www.linkedin.com/pulse/gms-throughput-improvement-process-kevin-kohls/. Accessed June 2022.

Lindberg, C.-F., Tan, S., Yan, J., and Starfelt, F. 2015. Key performance indicators improve industrial performance. In The 7th International Conference on Applied Energy, 28–31 March 2015, in Abu Dhabi, United Arab Emirates.

Luozzo, S., Varisco, M., and Schiraldi, M.M. 2020. The diffusion of international standards on managerial practices. International Journal of Engineering Business Management 12. https://doi .org/10.1177/1847979020921611.

Mohammed, A.R.K. 2020. A holistic approach for selecting appropriate manufacturing shop floor KPIs. PhD Thesis, University of Warwick. http://wrap.warwick.ac.uk/157871.

Muchiri, P. and Pintelon, L. 2008. Performance measurement using overall equipment effectiveness (OEE): Literature review and practical application discussion. International Journal of Production Research, 46(13), pp. 3517–3535. https://doi.org/10.1080/00207540601142645.

Muthiah, K.M.N. and Huang, S.H. 2007. Overall throughput effectiveness (OTE) metric for factory-level performance monitoring and bottleneck detection. International Journal of Production Research, 45(20), pp. 4753–4769. https://doi.org/10.1080/00207540600786731.

Myrelid, A. and Olhager, J., 2019. Hybrid manufacturing accounting in mixed process environments: A methodology and a case study. International Journal of Production Economics, 210, pp. 137–144.

Nakajima, S. 1988. Introduction to total productive maintenance, ISBN 13: 9780915299232, Productivity Press, Cambridge, MA.

Raju, S., Kamble, H.A., Srinivasaiah, R., and Swamy, D.R. 2022. Anatomization of the overall equipment effectiveness (OEE) for various machines in a tool and die shop. Journal of Intelligent Manufacturing and Special Equipment, 3(1), pp. 97–105. https://doi.org/10.1108/JIMSE-01-2022-0004

Raouf, A. 1994. Improving capital productivity through maintenance. International Journal of Operations and Production Management, 14(7), pp. 44–52. https://doi.org/10.1108/ 01443579410062167.

Sakai, H. and Li, J. 2020. New operational availability model to evaluate manufacturing throughput: Advanced TPS for global production. Universal Journal of Management, 8(3), pp. 96–102. https://doi .org/10.13189/ujm.2020.080306.

Shiau, Y.R. and Wang, S.Y. 2021. Key improvement decision analysis mechanism based on overall loss of a production system. Journal of Industrial and Production Engineering, 38(1), pp. 66–73.

Tang, H. 2019. A new method of bottleneck analysis for manufacturing systems. Manufacturing Letters (ISSN: 2213-8463) 19, pp. 21–24.

Tang, H. 2022. Quality Planning and Assurance—Principles, Approaches, and Methods for Product and Service Development, ISBN-13: 978-1119819271, Wiley, Hoboken, NJ.

Tang, H. 2023. OEE Review and Compatible Weighting Approach for OEE. In The 2nd Conference on Performance Managemente, Università di Roma Tor Vergata, Italy, 10 November, 2023.

Tsarouhas, P.H. 2020. Overall equipment effectiveness (OEE) evaluation for an automated ice cream production line: A case study. International Journal of Productivity and Performance Management, 69(5), pp. 1009–1032. https://doi.org/10.1108/IJPPM-03-2019-0126

Varisco, M., Johnsson, C., and Schiraldi, M.M. 2018. Proposal for a classification of ISO22400 KPIs for manufacturing operations management. In 23rd Summer School "Francesco Turco" —Industrial Systems Engineering, Vol. 2018. AIDI-Italian Association of Industrial Operations Professors. pp. 444–449.

Weiss, B.A., Horst, J. and Proctor, F. 2013. Assessment of Real-Time Factory Performance Through the Application of Multi-Relationship Evaluation Design NISTIR 7911, Department of Commerce. https://doi.org/10.6028/NIST.IR.7911.

Wudhikarn, R. 2010. Overall Weighting Equipment Effectiveness, Proceedings of the 2010 IEEE IEEM. In 2010 IEEE International Conference on Industrial Engineering and Engineering Management, Macao, China. 7–10 December 2010, National Institute of Standards and Technology, U.S. pp. 23–27. https://doi.org/10.1109/IEEM.2010.5674418.

Zhu, L., Johnsson, C., Varisco, M. and Schiraldi, M.M. 2018. Key performance indicators for manufacturing operations management–gap analysis between process industrial needs and ISO 22400 standard. Procedia Manufacturing, 25, pp. 82–88.

3

Bottleneck Identification and Buffer Analysis

3.1 Understanding of Bottleneck

3.1.1 System Bottleneck

3.1.1.1 Concept of Bottleneck

A bottleneck of a system is an element that constrains the system's performance. A pipeline can be analogous to a serial manufacturing system, where the narrowest section of a pipeline determines its flow rate. Figure 3.1 illustrates that the flow capacity of a pipeline depends on its narrowest section, creating a bottleneck that limits overall flow.

A bottleneck can be called a leverage point. From a resources perspective, the system's bottleneck is the subsystem with the highest resource utilization in a period. The resources can include equipment, time, workforce, and technology. Moreover, a system can experience multiple bottlenecks simultaneously.

Thus, it is essential to identify and address bottlenecks in a system for effective throughput management. While modifying any part of a system may impact its performance, improving the bottlenecks or leverage points can lead to substantial enhancements in the system's overall performance.

Gutenberg initiated the bottleneck concept more than 70 years ago [1951], and Kelley and Walker developed the critical path method (CPM) [1959]. A dedicated book on bottlenecks and their impacts on system performance is *"The Goal"* by Eliyahu M. Goldratt and Jeff Cox [1992]. The book introduced the Theory of Constraints (TOC) as a systematic approach to enhance system performance through bottleneck analysis and improvement. Figure 3.2 depicts the TOC process flow, concentrating on bottlenecks (or constraints). The five-step TOC process is a foundation of system performance improvement, which will be discussed further in Chapter 6.

Since the creation of the TOC, bottleneck analysis has become a popular topic and has been implemented in various types of business operations. Manufacturing operations, being visible and tangible, can easily adopt the principle of the TOC to systematically address system performance [Urban 2019; Myrelid and Olhager 2019; Naranje and Sarkar 2019; and Kuo et al. 2021]. In the late 1980s, General Motors (GM) employed a bottleneck identification process, leading to a significant improvement in manufacturing throughput. By 1997, productivity had increased by 26% [Kohls 2020]. Since 2004, the throughput improvement process has become a standard process at GM.

Manufacturing System Throughput Excellence: Analysis, Improvement, and Design, First Edition. Herman Tang.
© 2024 John Wiley & Sons, Inc. Published 2024 by John Wiley & Sons, Inc.
Companion website: www.wiley.com/go/Tang/ManufacturingSystem

Figure 3.1 Pipeline analogy of bottleneck limiting overall flow.

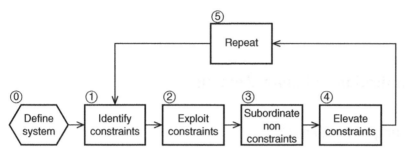

Figure 3.2 Five steps of applying Theory of Constraints.

3.1.1.2 Improvement on Bottleneck

As discussed in Chapters 1 and 2, the performance of manufacturing systems can be measured in terms of quantity, quality, efficiency, cost, etc. As a result, bottlenecks may be related to various issues, including the following:

- Throughput
- Quality concerns
- Cost
- Labor productivity
- Resource utilization
- Environmental concerns

Other factors that can impact performance include system design, budget, technology applications, and operational management. This book focuses on the bottlenecks associated with manufacturing processes and equipment, although the bottleneck-focused methodology can be applied to other aspects of a system.

In the context of manufacturing systems, addressing bottlenecks requires a targeted approach. Figure 3.3 provides an example of a manufacturing system composed of ten workstations. In this system, Workstation 2 has the lowest KPI score, making it the bottleneck. Hence, any improvement effort should focus on Workstation 2 to increase the system's throughput. Once improvements are made to Workstation 2, it is no longer the lowest performer, and Workstation 8 would become the new bottleneck in the system. Performance improvement efforts continue on the new bottlenecks identified.

While addressing bottlenecks one by one is sensible for simple systems, the dynamic nature of complex systems presents unique challenges. For instance, bottlenecks can be at multiple locations at the same time, with multiple contributing factors, and changing over time. This challenge requires that bottlenecks be collectively and continuously addressed as a set, considering their impacts and interactions with various means and methods to deal with different root causes.

In addressing bottlenecks, temporary actions, such as working overtime, may offer immediate relief and increase throughput for a moment. However, temporary actions are not cost-effective in the long run and are unlikely to be sustainable, because they do not address the root causes. Therefore, it is essential to identify and resolve the root causes of bottlenecks for lasting improvement.

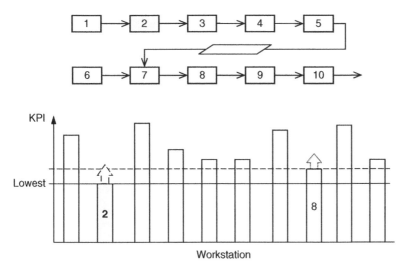

Figure 3.3 Example of bottlenecks in a manufacturing system.

Bottleneck identification and elimination are a process of continuous improvement for operational management on the shop floor. As a unique methodology, a bottleneck-focused approach should be jointly with other problem-solving and continuous improvement methods, including PDCA, A3, DMAIC, and 8D/5D. Further discussion on the integration of TOC and other methods will be explored in depth in Chapter 6.

3.1.1.3 Considerations in Addressing Bottleneck

While bottlenecks are often noticeable in manufacturing, they may be in other forms, including policies, communication, or skills. A study of three companies found that 64% of bottlenecks were related to production, while the remaining 36% were attributed to other factors [Schultheiss and Kreutzfeldt 2009]. The workload, communication failure, material unavailability, and capacity deficit were 36%, 26%, 19%, and 17% of the reasons, respectively, for the bottlenecks.

To facilitate improvement project work and communications, it is recommended that physical elements or entities in a system be referred to as bottlenecks. In many cases, the associated subordinate and intangible constraints may be referred to as reasons or root causes of the physical operation's constraints. For instance, in the system shown in Figure 3.3, Workstation 2 is the bottleneck. The bottleneck may be caused by equipment, worker skills, incoming materials, and so on.

Bottlenecks can vary in location and duration because of various reasons, including operational changes, improvement efforts, and the randomness of equipment downtime. When working on throughput improvement over an extended period, such as a month, the likelihood of bottlenecks remaining in the same locations can be high. Therefore, when conducting throughput improvement projects, it is recommended to focus on consistent bottlenecks over a week. The classification of bottlenecks is further discussed in the next subsection 3.1.2.

While addressing bottlenecks is crucial, designing systems to avoid bottlenecks altogether can be more effective. Bottleneck avoidance is an important task for system design, which will be discussed in Chapter 7. Ideally, a system can be designed with no inherent bottlenecks, meaning all functions can run smoothly by design. To illustrate the importance of balanced design, consider a personal computer. It comprises several key elements such as CPU, GPU, SSD, memory, and I/O devices. All these components should operate at the same performance level. An element that is either overpowered or underpowered can cause waste or bottlenecks in computer performance, respectively.

In the real world, some bottlenecks of manufacturing systems are design-induced, called capacity bottlenecks to be discussed in the next subsection 3.1.2.1. When design-related bottleneck root causes are identified, operational management needs to work with design teams to address the root causes.

It is crucial to emphasize that while bottlenecks are pivotal to system throughput, but they are not the sole concern. Other critical aspects of operations, such as employee safety and regulatory compliance, hold even greater importance. These subjects, although crucial, are not the focus of this book.

Subsection Summary: Understanding and addressing bottlenecks are vital for improving system performance. This subsection explores the concept and impact of bottlenecks, along with the considerations for addressing them.

3.1.2 Classification of Bottleneck

3.1.2.1 Performance and Capacity Bottlenecks

Bottlenecks can be categorized in several ways. From a root cause perspective, there are two basic reasons for bottlenecks in a system:

- Performance bottlenecks: These are typically related to operational management, including maintenance management and quality management. This type of bottleneck can be caused by equipment failures, operational downtime, defective parts, high variation, and so on. They can be temporary and change their locations. Operational management should focus on this type of bottleneck and the corresponding root causes.
- Capacity bottlenecks: These are primarily related to system and process design. The weakest link, or least capable element, of a manufacturing system is often a result of design. This type of bottleneck is typically consistent in their locations over time. Operational management may reduce the impacts of this type of bottleneck but is not able to eliminate them, without modifying the process or system design. The most effective solution is to redesign or modify the process/system.

Figure 3.4 summarizes both performance bottlenecks and capacity bottlenecks with the main elements, displaying their complex relationships with various aspects.

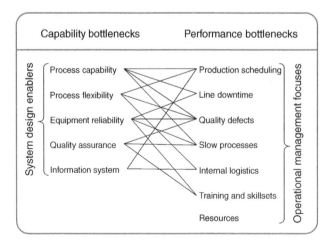

Figure 3.4 Two basic types of throughput bottlenecks and their relations.

Throughput problems on the production floor can be complex, resulting from multiple factors related to both operational performance and design capacity. Understanding throughput issues and their root causes is essential for effectively enhancing system throughput performance. If a root cause is related to system design, manufacturing practitioners can analyze the issue and give feedback to design professionals for system design change. Simultaneously, operational management can implement temporary measures to maintain system throughput.

As shown in Figure 3.4, performance bottlenecks and corresponding throughput improvement projects can involve various aspects or departments in a plant. Therefore, a throughput improvement team should consist of cross-functional professionals who collaborate to address various root causes and propose solutions. Once root causes are identified, the relevant department can take the lead in executing the solutions.

3.1.2.2 Time Characteristics of Bottlenecks

As previously discussed, bottlenecks in a system are caused by various random and nonrandom factors, making them dynamic. As a result, bottlenecks can change location, duration, and frequency over time. Analyzing bottleneck locations over different time periods, such as hour, day, week, and month, can yield different results, which can be insightful for understanding the bottlenecks and for problem-solving.

Considering the time characteristics of bottlenecks, they can be categorized into the following three types: momentary, medium-term, and persistent. The classification of these three types depends on production volume, and their boundaries can be approximate and overlap.

- Momentary bottlenecks are short-lived, typically lasting several minutes but less than a couple of hours, in the context of mass production. They are often caused by random factors in operations and associated with the routine work of operational and maintenance management.
- Persistent bottlenecks stay for weeks or even longer or recurrently occur over an extended period, even with improvement efforts. They are primarily caused by system design issues such as an unreliable process or equipment.
- Medium-term bottlenecks fall in between, lasting from a few hours to a week, or recurring periodically. Their challenges in finding solutions may arise due to the combination of operational management and system design issues.

Figure 3.5 illustrates an example of the persistence and general contributors of three types of system bottlenecks in mass production. Identifying their types can help in understanding the reasons and effective solutions to resolve them and prevent recurrence. These time-based bottleneck categories can serve as a useful reference for maintenance management.

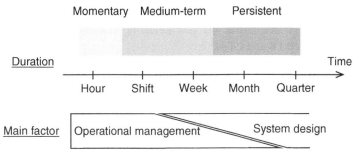

Figure 3.5 Duration and persistence of bottlenecks and their general contributors.

Usually, it is effective but challenging to deal with persistent bottlenecks, as some design changes can be needed. Concentrating resources on addressing these persistent and recurring issues with common causes can enhance manufacturing performance both now and in the future.

Subsection Summary: Bottlenecks can be classified into types of capacity and performance, primarily related to system design and operational management, respectively. Their durations are also associated with different root causes. Understanding these classifications helps in effectively identifying and addressing bottlenecks.

3.2 Bottleneck Identification

The first crucial step in throughput improvement is to identify both what and where the bottlenecks are. This initial identification is fundamental to subsequent analysis and problem-solving tasks. Various KPIs can be employed for pinpointing bottlenecks, provided they are deemed critical to system performance technically and financially.

3.2.1 Identification using Individual KPI

3.2.1.1 By Operational Availability

Manufacturing systems ideally run smoothly without downtime and waiting time. The system's operational availability (A) and standalone availability (A_{sa}) can be calculated with known the operational states of a system, refer to subsection 1.4.2.

$$A_{sa} = \frac{\text{Actual production time}}{\text{Planned production time} - \text{Starved time} - \text{Blocked time}} (\%)$$

Referring to the standalone concept in Chapter 1, A_{sa} reflects a true throughput performance, being a reliable KPI, for bottleneck identification, as it excludes the influence of external factors.

Here is an example to illustrate the application of standalone availability in identifying bottlenecks. A system in a serial setting with four operations; each operation has 95.3%, 92.6%, 98.1%, and 94.3% of A_{sa}, respectively. In this case, the second operation is the bottleneck because it has the lowest A_{sa}. It is important to reiterate that using overall operational availability (A) can be misleading for bottleneck identification, because it does not consider the starved and blocked times. The significance of utilizing standalone KPIs for bottleneck identification will be further explored in the following subsections.

3.2.1.2 By Throughput Rate and Cycle Time

In addition, throughput rate (TR) is another key focus in throughput management. When downtime and waiting (starved and blocked) time are not major concerns, cycle time (CT), which inversely affects TR, can be considered for bottleneck analysis. Referring to subsection 1.2.2 of Chapter 1, the overall relationship between TR and CT is:

$$TR = \frac{3600}{CT \text{ (seconds)}} \text{ (JPH)}$$

Figure 3.6a illustrates a simple example of a system with four operations. Figure 3.6b shows the system's performance, in terms of CT and TR.

In a sequential system like this, the system's TR is limited by the slowest operation. In this example, Operation 2 is the slowest, constraining the system's performance to around 42 JPH. Sometimes, ranking a KPI, such as TR, in a Pareto chart helps quickly identify the bottleneck.

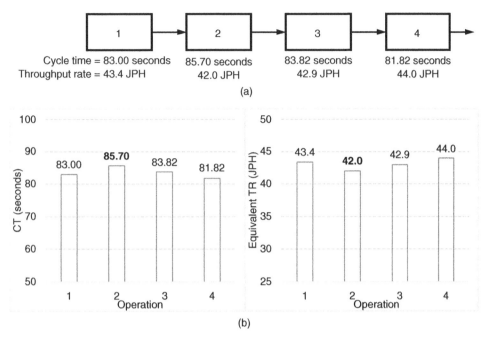

Figure 3.6 Example of a system with four operations with their performance.

Sometimes, using TR data can be more straightforward for interpretation than using CT data. While measuring actual CT can reveal deviations in performance from the design CT. In addition, significant variations in CT measurements can be a throughput concern and may warrant further analysis to identify their root causes.

While using TR and CT data is valuable for bottleneck identification, it is important to note that this approach is valid only when there is no significant waiting time. If this condition is met, the exclusion of starvation and blockage or the application of the standalone concept becomes crucial for accurately pinpointing system bottlenecks.

3.2.1.3 By Active Period

Bottlenecks can be identified by analyzing resource utilization, as bottlenecks experience the highest utilization during a period. Thus, bottlenecks can be identified based on active or busy time.

Active time refers to the time during which the system is engaged in scheduled work, including working, maintenance, setup, changeover, and unscheduled downtime. On the contrary, inactive time represents the system's waiting time, being blocked and/or starved, as shown in Figure 3.7.

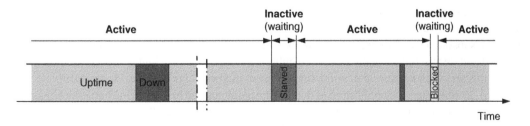

Figure 3.7 Active and inactive periods during manufacturing operations.

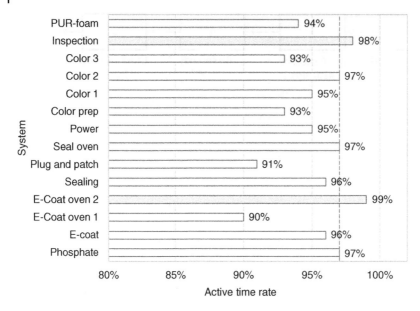

Figure 3.8 Bar chart of system active time rates.

In a system with multiple operations, the bottleneck is identified as the operation with the longest continuous active period in a given timeframe. This approach is sometimes referred to as the active period method [Roser et al. 2001; Li et al. 2009].

One can assess the active rates of all operations or equipment in a system over a period, such as a week. The active rate shows how busy a system is, represented as a percentage:

$$\text{Active rate} = \frac{\text{Active time}}{\text{Total time}} \, (\%)$$

where the active time is defined above. The total time includes active time and inactive time.

Figure 3.8 shows an example of the active rates in a vehicle paint shop. If an active rate threshold is 97%, then systems E-Coat Oven 2 and Inspection are identified as the primary bottlenecks because their active rates are very high, surpassing the threshold.

Active time analysis proves to be an effective way to identify system bottlenecks, considering both internal and external factors. A study by Toyota indicated that using the average length of all active periods was more effective for locating bottlenecks than using the percentage of the total active time based on computer simulation [Roser et al. 2001].

3.2.1.4 By Starved and Blocked Time

Like the active period method, analyzing the inactive time, or starved and blocked time, of subsystems can help identify system bottlenecks. A bottleneck tends to cause blockage and starvation in other subsystems and operations, while bottlenecks themselves exhibit the lowest combined blocked and starved time. Because all the neighboring subsystems have higher blocked and starved times, this approach is referred to as the turning point method [Li et al. 2009; Plomp 2019].

Figure 3.9 illustrates the application of the turning point method in identifying bottlenecks based on the analysis of starved and blocked time in a system comprising 20 machines arranged in a serial setting [Wang et al. 2021]. Figure 3.9a is a stacked chart; Figure 3.9b is a clustered bar chart. Machine 14, the bottleneck, is positioned in the valley between the peaks of blocked and starved time in both chart formats.

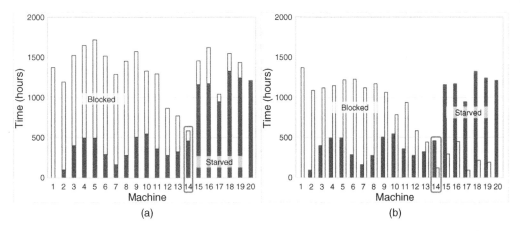

Figure 3.9 Bottleneck identification based on turning point (Wang et al. 2021/with permission of Springer Nature).

Subsection Summary: Several bottleneck identification methods based on individual KPIs include operational availability, throughput rate, cycle time, active period, and turning point. Applying these KPIs and techniques is straightforward and provides a data-driven process to locate system constraints.

3.2.2 Identification using Multiple KPIs

3.2.2.1 Examining Multiple Metrics

While individual KPIs offer insights into specific aspects of manufacturing system performance, combining multiple KPIs may provide a comprehensive analysis of bottlenecks. Particularly for complex systems, employing multiple KPIs can identify different types of bottlenecks and find the corresponding root causes.

Figure 3.10 illustrates an example of GM's work on throughput improvement [Alden et al. 2006] on an assembly line. It consists of the following eight workstations: Load, Weld, Inspect, Robogate, Weld 1, Weld 2, Seal, and Unload. GM used four KPIs to assess a vehicle body assembly line: TR (measured in JPH), mean time between failures (MTBFs, in cycle), mean time to repair (MTTR, in minutes) or downtime, and quality scrap ratio (in percentage).

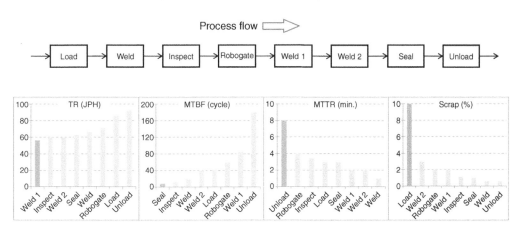

Figure 3.10 Example of production line with different KPIs (Adapted from Alden et al. 2006).

In this example, when considering the throughput rate, Weld 1 is the bottleneck due to its lowest output. To improve the system's operational availability (in the form of MTBF), Seal requires attention because it has the lowest reliability or the shortest MTBF. Likewise, Unload operation warrants attention because of its extended repair time (MTTR), a critical concern. A more detailed discussion of MTBF and MTTR will be presented in Chapter 5.

As illustrated in the example, addressing multiple KPIs can be complex, leading to conflicts in identifying bottlenecks. Recognizing the complexities introduced by multiple KPIs, the need for prioritization becomes evident. Addressing the most critical concerns or those with the highest cost implications is essential. For example, if reducing quality scrap is the main concern, prioritizing it over other KPIs can be reasonable.

Furthermore, bottlenecks may interrelate with each other or possess confounding effects, posing challenging analytical tasks that require additional effort. For example, quality issues themselves are worrisome and their impacts on downstream operations and final products can be even more impactful. The interplay among various aspects of a manufacturing system and their potential confounding effects will be further explored in the upcoming chapters when touching on this topic.

3.2.2.2 Using OEE and Standalone OEE

In addition to using multiple KPIs, using a composite KPI embedding individual KPIs is a beneficial method. As discussed in Chapter 2, overall equipment effectiveness (OEE) is a comprehensive indicator consisting of three element KPIs: operational availability (A), speed performance (P), and quality rate (Q). Thus, OEE encompasses the three major performance aspects of a system, indicating overall performance.

Calculated from the data of the GM example in the previous subsection 3.2.2.1, OEE can serve as a composite KPI for bottleneck identification. In this stance, the operational availability is derived from the MTBF and MTTR, refer to Chapter 5 for the formula. The performance is calculated with a design TR assumption of 70 JPH. The quality rate is determined by converting the scrap rate. Figure 3.11 displays the calculated OEE values for the eight workstations based on the same data set. Based on this OEE analysis, Seal workstation is the system bottleneck, which aligns with the findings from the MTBF analysis (Figure 3.10).

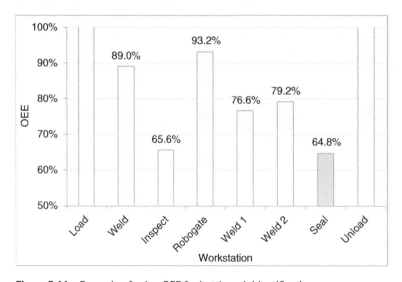

Figure 3.11 Example of using OEE for bottleneck identification.

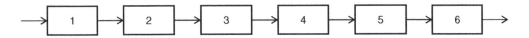

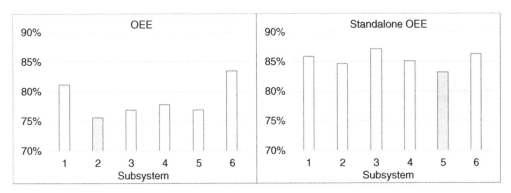

Figure 3.12 Using OEE and standalone OEE for bottleneck identification.

When using OEE, TR, and other KPIs for bottleneck identification, it is also crucial to consider the standalone concept to isolate external influences like starvation and blockage (discussed in Chapter 1). Figure 3.12 shows an example of a system's OEE and standalone OEE application for bottleneck identification.

If considering OEE, Subsystem 2 is a bottleneck due to its lowest OEE level. However, based on the standalone OEE values, excluding external influences, Subsystem 5 has the lowest standalone performance, making it the true bottleneck. By comparing overall and standalone OEE metrics for each subsystem, issues can be accurately pinpointed.

3.2.2.3   Considerations in Bottleneck Identification

These KPIs and methods, including single KPI, multiple KPIs, composite KPIs, have their own benefits and challenges. For instance, the active period method may require more data for identifying bottlenecks than a method based on downtime information. Using cycle time requires effort in measuring and verifying CT.

Moreover, these methods are straightforward to apply to serial systems, as discussed earlier. In the case of complex systems with diverse local configurations, such as parallel, hybrid, and loop branches, additional analysis is needed on the system configurations to make these methods applicable. System configuration analysis is discussed in Chapter 6, as exemplified by Figure 6.22.

These various methods discussed are general. For specific manufacturing systems, other KPIs and respective analyses may be more suitable. For instance, in battery manufacturing, bottleneck identification and reduction are based on factors, such as energy demands and process time, as studied by Silva et al. [2021].

It is important to iterate that identifying bottlenecks, using these and other methods, is the first step for throughput improvement. Equally crucial is finding the root causes of the identified bottlenecks. From this perspective, analyzing various KPIs can be helpful. For instance, active time encompasses both scheduled work time and unscheduled downtime. The former is influenced by system design (discussed further in Chapter 8) and production scheduling. The latter is linked to system reliability, a topic thoroughly explored in Chapters 5 and 7.

Subsection Summary: This subsection explores the use of multiple and composite KPIs, along with respective analyses, for identifying bottlenecks. The characteristics of examining multiple performance metrics and standalone applications are discussed, along with examples. Leveraging multiple and composite KPIs can provide systematic insights into bottleneck identification.

3.2.3 Research on Bottleneck

Bottleneck identification in complex systems can be challenging for manufacturing practitioners. Researchers have been working on the topic for over the past two decades.

3.2.3.1 Challenges of Bottleneck Identification

The concept of bottlenecks and bottleneck identification are straightforward for small systems in a serial setting. However, the complexity of large systems introduces challenges in the identification process. Figure 3.13 provides an example of a vehicle paint system layout, where the conveyors (buffers) are represented by parallelograms labeled starting with "B" [Tang 2018]. Conveyors function as buffers between subsystems, absorbing small throughput issues and variations and alleviating the impact of bottlenecks. Therefore, some bottlenecks may not be easy to identify. The analysis of buffers will be thoroughly discussed in Section 3.3.

In addition, bottlenecks in complex systems become dynamic during manufacturing operations. Real-time interactions between subsystems can result in blockage and starvation in the subsystems. Interactions with buffer effects can cause bottlenecks to change in location and duration. Thus, buffer identification in complex systems requires further study.

These challenges in identifying bottlenecks give rise to a crucial aspect: bottleneck shifts, referring to the changes in the location of bottlenecks in a system. The shift results from the working states of connected subsystems and interactions with workstations and buffers. For instance, a workstation may no longer be the bottleneck if it becomes starved or blocked, causing the bottleneck to relocate elsewhere in the system. This phenomenon is demonstrated in a simulation study conducted by Roser et al. [2017].

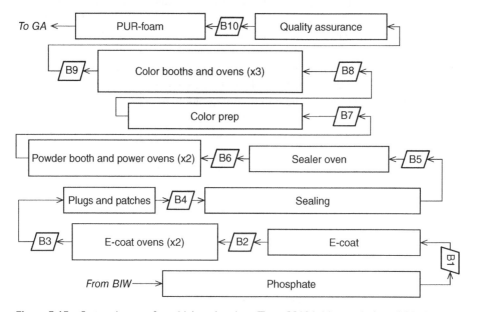

Figure 3.13 System layout of a vehicle paint shop (Tang, 2018/with permission of SAE International).

To address the dynamic nature of bottlenecks, it is advisable to focus on the persistent ones, as discussed earlier. In addition, simultaneously addressing multiple bottlenecks, e.g., the top three, is another effective strategy. Further discussion on this topic is available in Chapter 6.

The challenges associated with bottleneck identification drive academic research toward the development of new methodologies, a discussion that will continue in the next subsection 3.2.3.2.

3.2.3.2 Research Methodology on Bottleneck

In academic research, the methods of bottleneck identification can be categorized into the following two groups: computer simulation and mathematical modeling. These studies offer insights for practical benefit bottleneck detection, and some findings may be implemented on the production floor through research advancements.

Mathematical modeling is an interesting approach explored for bottleneck identification to enhance the understanding of bottlenecks and their characteristics. However, a challenge in mathematical modeling is that established models are often based on simplified conditions and assumptions. These simplifications can limit the applicability of mathematical models. As a result, most modeling methods remain theoretical with limited validation and practical applications. Further development of mathematical modeling should make the models closer to real-world manufacturing conditions by involving improved collaboration between academic researchers and experienced manufacturing practitioners.

To detect momentary bottlenecks, one method involves measuring the waiting time of the parts or the length of the queue in front of a machine or system [Lawrence and Buss 1994; Betterton and Silver 2012]. In addition, a few studies investigated the active period method, as discussed earlier. Real-time bottleneck analysis has also been studied, where the loss of value added by idle machines is analyzed while considering the relationship between machines and operators [Wedel et al. 2015]. From a throughput improvement perspective, detecting momentary bottlenecks has limited practical value in most real-world situations, as these bottlenecks are not typically the primary focus.

Research on bottlenecks also includes predicting them based on existing data. For example, an autoregressive moving average (ARMA) model was developed to predict near-future bottlenecks by transforming the blocked and starved times of each workstation into time series data [Li et al. 2011]. Bottleneck prediction can be meaningful if the bottlenecks predicted have major impacts on manufacturing operations and throughput.

In manufacturing research, computer discrete-event simulation (DES) is a rising star as an effective method for identifying bottlenecks and has been studied by various researchers [Mourtzis 2020]. Particularly, integrating with other methods, DES can play a significant role in throughput management. An approach combined statistics and simulation methods to make a prediction [Subramaniyan et al. 2018, 2019]. Another study integrated the simulation model with bottleneck detection methods for a pharmaceutical production process [Eskandari et al. 2013]. A study proposed using a system dynamics-based simulation model with throughput accounting to investigate product-mix problems that arise under stochastic demand and scrap ratio [Hilmola and Gupta 2015].

Many large corporations utilize DES in developing new manufacturing systems. Simulation results are becoming more accurate and can closely reflect real-world scenarios. However, there are still barriers to wider applications of simulation methods in industry, due to the cost of software and the need for specialized skills. Chapter 8 delves deeper into the applications and challenges of DES.

3.2.3.3 Bottleneck Analysis for Nonmanufacturing

The application of bottleneck analysis and identification extends beyond manufacturing operations, encompassing areas such as supply chain management, logistics, and product design. For example, studies have demonstrated the throughput performance of a production system as a function of the existing order backlog, illustrated as a logistic operating curve (LOC) [Schultheiss and Kreutzfeldt 2009; Mütze et al. 2023]. Furthermore, researchers have published studies on bottleneck considerations in engineering fields [Hinckeldeyn et al. 2014; Mehrbod and Tory 2020].

In addition, bottleneck analysis is crucial in other types of operations, such as healthcare and emergency services. A study example is on head computed tomography for emergency department (ED) patients. This study revealed that the process bottleneck varied for all 8312 ED patients, shifting among four steps: physician order, scheduling, completion, and report [Rogg et al. 2017]. Each process served as the bottleneck for 30%, 1%, 27%, and 42% of the time, respectively. This bottleneck information can be valuable for operational management.

In a study on the microfinance sector and financial inclusion in a low-population-density, low-income country, researchers analyzed secondary data and conducted expert interviews to understand the bottleneck situation and its root causes [Bouasria et al. 2020]. In agroecology, another study analyzed sustainable agroecological farm reproduction and identified the primary bottleneck as the closure of biogeochemical cycles within agroecology territories [Padró et al. 2020]. These studies demonstrate the broad relevance of bottleneck analysis outside of traditional manufacturing and production environments.

Cross-disciplinary studies offer valuable insight and perspective for developing novel methods for bottleneck identification in manufacturing, and vice versa. The exchange of knowledge and experience across different fields can enhance the understanding and effectiveness of bottleneck analysis in various domains.

Subsection Summary: Identifying bottlenecks can be challenging due to system complexity. Research sheds light on potential approaches for the applications on the production floors and other operations beyond manufacturing.

3.3 Understanding of System Buffer

A buffer is a cushion against the shock of fluctuations in business activity, as defined by Merriam-Webster [n.d.]. In the context of a manufacturing system, a buffer mechanism often presents itself as a resource, taking different forms. A buffer can be work in process (WIP) units or inventory, excessive process capacity, extra time or delay, or a combination of these to compensate for variations and minor interruptions in a system. In this section, the characteristics, status, and effects of WIP in buffers are studied, with further discussion on buffer design in Chapter 8.

3.3.1 Buffer Effect of Conveyor

3.3.1.1 Function of Conveyor

Conveyor systems play an integral role in manufacturing, facilitating the transfer of parts and WIP units between subsystems. Figure 3.14 illustrates a typical conveyance process in a system, including the following three steps:

1. Unloading WIP units from subsystem A to the conveyor, taking time "t_{unload}"
2. Transferring the WIP units along the conveyor, taking time "$t_{transfer}$"
3. Loading the WIP units to subsystem B from a conveyor, taking time "t_{load}"

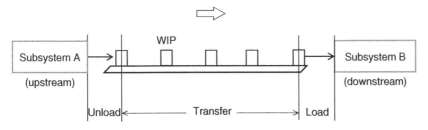

Figure 3.14 Process steps of conveyor (buffer) function.

In many instances, the tasks and time of the unload and load functions are incorporated into subsystems, instead of being part of the conveyance process. Accordingly, the time analysis of a conveyor (buffer) is simplified.

In addition to the transfer time, another key parameter of a conveyor is the number of WIP units during the conveyance process. Consequently, the WIP units on conveyors can be treated as internal stock, making the use of a conveyor as a buffer a natural choice. Therefore, when designing a conveyor, it is essential to ensure that it can accommodate the desired number of WIP units. Understanding the buffering function of a conveyor, along with its time and WIP unit attribution, is crucial for optimizing the throughput performance of a manufacturing system.

3.3.1.2 Understanding of Buffer Effect

It is about time to shift focus from conveyance functions to assessing the conveyor's impacts on manufacturing operations. In the system depicted in Figure 3.14, when Subsystem A is down, the WIP units on the conveyor can keep Subsystem B running until the conveyor is empty. The conveyor's buffer effect reduces the starved time of Subsystem B from Subsystem A's downtime.

Similarly, if Subsystem B is down, the empty space on the conveyor can keep Subsystem A running until the conveyor is full. In this scenario, the conveyor/buffer reduces the blocked time of Subsystem A.

The analysis of the two scenarios suggests that the buffering effect's extent is related to the buffer size, that is, the number of WIP units and empty space. If a buffer were infinitely large, allowing for an unlimited number of WIP units and unlimited space on a conveyor, the subsystems could have operated fully independently. This indicates a conveyor size matters to operational performance.

Therefore, a conveyor can partially decouple the two subsystems connected. Figure 3.15 summarizes a generalized conveyor's buffering effects. A conveyor can reduce, not eliminate, the propagation of downtime losses. With the buffer effects, a conveyor system can enhance the efficiency and resilience of manufacturing operations by reducing starved and blocked time.

In volume production, specialized processes and equipment, such as resequencing mechanisms, can also function as buffers. WIP resequencing involves organizing a random queue of WIP units into groups based on certain attributes, such as size, color, or components, to optimize

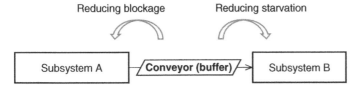

Figure 3.15 Buffering effects of a conveyor on connected systems.

manufacturing processes. A resequencing area, which usually can handle many WIP units, provides good buffering to improve operational performance.

The next subsection 3.3.1.3 discusses the quantitative buffer effect to enhance understanding. It is crucial to recognize that the buffer effect is tied to the actual real-time performance of the two connected systems (e.g., A and B in Figure 3.15) and the interactions between the buffer and these systems. The actual buffer effect can be accurately predicted through computer simulations, as detailed in Chapter 8.

3.3.1.3   Discussion of Buffer Effects

Buffer effects can be reviewed from a different angle. In a manufacturing system, there are often multiple conveyors in place. A subsystem typically connects two conveyors: an upstream conveyor to supply parts or WIP units, and a downstream conveyor to move the finished units to the next subsystem. Figure 3.16 illustrates the effects of upstream and downstream buffers, resulting in an improved TR.

In this scenario, the effects of buffers on the operations of Subsystem O can be summarized as follows:

- When the upstream subsystem stops working, the WIP units in Buffer A continue to feed Subsystem O until Buffer A is empty, thereby reducing Subsystem O's starved time.
- When the downstream subsystem stops working, the empty space in Buffer B continues to receive finished units from Subsystem O until Buffer B is full, thus reducing Subsystem O's blocked time.
- The buffering effects on reducing starved and blocked times are typically limited by the number of WIP units and empty spaces in the buffer. In mass production, the reduction is only a matter of a few minutes.

In an ideal scenario, the number of WIP units on a conveyor is at approximately 50% of its capacity. For instance, if a conveyor can hold 20 WIP units, a consistent count of ten units is deemed ideal.

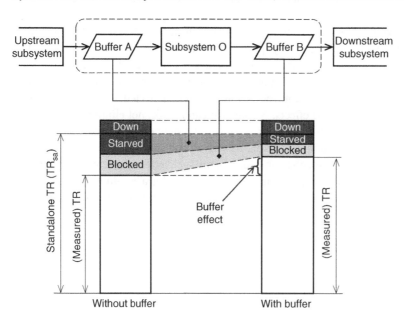

Figure 3.16   Effects of upstream and downstream buffers on system throughput.

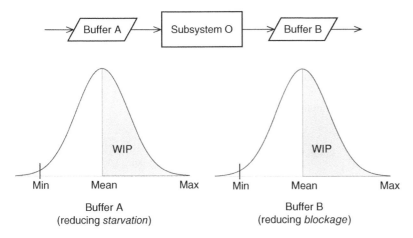

Figure 3.17 WIP distribution in a buffer for buffer effect discussion.

In addition, it can be assumed that the WIP units in a buffer follow a normal distribution, which is approximated from a Poisson distribution. This distribution is depicted in Figure 3.17 where the shaded area represents the WIP units.

Subsection Summary: Conveyors in manufacturing systems play a crucial buffering role, mitigating the propagation of downtime losses and boosting system throughput. Conveyor size and WIP distribution determine the buffering effect, which will be quantitatively analyzed in the next subsection 3.3.2.

3.3.2 Buffer Effect Analysis

3.3.2.1 Illustrative Effect Analysis

Expanding on the earlier discussion of the overall buffer effect, the buffer effect can be discussed quantitatively. When the size of a buffer is known, its effects on reducing starvation and blockage in a manufacturing system can be quantitatively estimated. For the sake of discussion, let us consider a simplified example. A conveyor has a capacity of ten units, with five WIP units and five-unit spaces in a normal situation. A WIP unit takes one minute to transfer on the conveyor. The CT for both Subsystems A and B is 60 seconds (refer to Figure 3.15).

The buffer effects in this case can be analyzed in multiple situations. When Subsystem A experiences a downtime of ten minutes, the five WIP units on the conveyor can sustain Subsystem B's operation for five minutes. When Subsystem A resumes operation, a new WIP unit will take a minute to reach Subsystem B. As a result, Subsystem B experiences starvation for a total of six minutes, benefiting four minutes from the buffer effect, listed as Situation 1 in Table 3.1. Without a buffer, Subsystem B would have experienced the same ten minutes of starvation due to Subsystem A's downtime.

In another situation, Situation 2 in Table 3.1, Subsystem B experiences a downtime of ten minutes. The vacant space equivalent to five WIP units on the conveyor will be filled by Subsystem A. As a result, the blocked time for Subsystem A is reduced by five minutes. However, there is no change in the throughput of the entire system at the end of Subsystem B, meaning the buffer does not contribute to the entire system throughput in this case. While the full conveyor can be helpful to Subsystem A's downtime if it immediately occurs after this event.

Table 3.1 Buffer effect analysis on two connected systems.

Situation	Subsystem A	Conveyor/buffer	Subsystem B	System throughput
0	Running	Running	Running	Normal (no buffer effect)
1	Down 10 minutes	Running	Operational, starved 6 minutes	Lost 6 minutes (4-min buffer effect)
2	Operational, blocked 5 minutes	Running	Down 10 minutes	Lost 10 minutes (no buffer effect)

This exemplary discussion qualitatively and quantitatively demonstrates the buffer effects of a conveyor. The example highlights that the buffer effect does not directly impact the entire system's throughput in situations caused by the downstream system's interruptions.

3.3.2.2 Consideration of Buffer Reliability

It is worth noting that conveyors themselves are not 100% reliable and may experience downtime as well. In such cases, the buffer effect can be analyzed comparably, and the results are presented in Table 3.2 as an extension of Table 3.1.

In Situation 3, both Subsystems A and B are operational, but the conveyor is down for two minutes. Both Subsystems A and B experience a two-minute blockage and starvation, respectively. The entire system loses two minutes in its production.

If Subsystem A is down for ten minutes, and then the conveyor is down for two minutes (Situation 4a), the total loss of system throughput is eight minutes. The loss is the combination of Situation 1 in Table 3.1 and the conveyor's 2-minute downtime. However, if both downtimes occur simultaneously (Situation 4b), the conveyor downtime may be masked by the longer Subsystem A's downtime. The varying situations, such as the same length of downtime of the conveyor and Subsystem A or the downtime of the conveyor being longer than Subsystem A's, can be analyzed in the same way but with different results. Readers may be interested in analyzing these situations.

Table 3.2 also lists two situations, 5a and 5b, where the downtimes of Subsystem B and the conveyor occur serially and concurrently. They can be analyzed similarly. As an extension of Situation 2 in Table 3.1, Situations 5a and 5b can be compared with Situation 2 for their differences.

Table 3.2 Influence of buffer reliability on throughput of two connected systems.

Situation	Subsystem A	Conveyor/buffer	Subsystem B	System throughput
3	Operational, blocked 2 minutes	Down 2 minutes	Operational, starved 2 minutes	Lost 2 minutes
4a	Down 10 minutes	Down 2 minutes (serially)	Operational, starved 8 minutes	Lost 8 minutes
4b	Down 10 minutes	Down 2 minutes (concurrently)	Operational, starved 6 minutes	Lost 6 minutes
5a	Operational, blocked 2 minutes	Down 2 minutes (serially)	Down 10 minutes	Lost 12 minutes
5b	Operational, blocked 2 minutes	Down 2 minutes (concurrently)	Down 10 minutes	Lost 10 minutes

The quantitative analysis above on the simplified cases implies the situations can be more complex in the real world. Additional situations can be numerous combinations and variations, including downtime variation, WIP variability in a buffer, and downtime relationship between the subsystems and conveyor. Therefore, analyzing buffer effects on system throughput performance can be too intricate for simple analysis. Computer simulations become effective tools for evaluating buffer effects within a specific manufacturing system.

Nevertheless, it can be concluded that buffers are effective in covering brief downtimes, small interruptions, and variations, thus aiding system throughput.

3.3.2.3 Characteristics of WIP on Buffer

To study the buffer effect, a buffer (conveyor) can be characterized by considering the number and distribution of WIP units within it. These factors significantly influence the effects of a buffer.

When analyzing the characteristics of WIP on a conveyor, it is reasonable to assume that WIP units are evenly positioned along the conveyor, as illustrated in Figure 3.14. Under this assumption, if there is only one WIP unit on a conveyor, it would be at the left of the conveyor and travel halfway down before reaching the downstream subsystem. Consequently, a single WIP unit does not effectively serve as a buffer.

A key question is how many WIP units must be on the conveyor to ensure a continuous and timely supply to the downstream subsystem. A conveyor system can be viewed as an operation between two connected subsystems. In an ideal scenario of one-piece continuous flow, the total transfer time should be the same as the cycle time of the connected subsystems to maintain a seamless operation. To account for variations, it is recommended to add one additional unit. Therefore, a simple method for estimating the minimum number of WIP units is:

$$\text{Min} = \frac{t_{\text{transfer}}}{\text{CT}} + 1 \text{ (units)}$$

where "Min" represents the minimum number of WIP units on a conveyor, t_{transfer} is the entire transfer time, and CT stands for the cycle time of the subsystems connected. Increasing the WIP travel speed on a conveyor can reduce t_{transfer}, provided it is technically feasible.

For example, if $t_{\text{transfer}} = 80$ seconds and CT $= 60$ seconds, the calculated Min is:

$$\text{Min} = \frac{80}{60} + 1 = 2.33 \text{ (units)}$$

It is advisable to round up the calculated value, considering variation from the assumption of the evenly positioned WIP on a conveyor. So, the minimum WIP is three units in this case.

Conversely, a conveyor has a limited size, determining its capacity to hold WIP units. This maximum capacity is referred to as "Max" and is determined by the system design. Therefore, for the buffer effect, a conveyor has two values of WIP units: Min and Max.

Subsection Summary: Considering the size and reliability of conveyors, a simple example is used to quantitatively examine the buffer's roles in mitigating interactions between connected subsystems. The buffer characteristics, including the number and distribution of WIP units, influence the buffer effects.

3.4 Buffer Analysis for Bottleneck Identification

The status and changes of WIP on conveyors/buffers also convey implicit messages regarding the operational status of a system and the presence of bottlenecks. Appropriate analyses can reveal such messages. This section aims to delve into buffer information to identify bottlenecks.

3.4.1 Considerations in Buffer Analysis

3.4.1.1 Buffer WIP Monitoring

In automated systems, WIP units on conveyors are often monitored and reported in real-time, providing crucial information for operational management. The status of WIP on conveyor systems serves as indicative data for system performance.

When upstream and downstream subsystems are operating normally, the number of WIP units in a buffer (conveyor) should remain stable. As discussed in the last subsection 3.3.2 on buffer effects, a buffer is ideally about half full to mitigate blockage for the upstream system and starvation for the downstream system. However, in actual production, buffer WIP levels can vary from empty to full.

As mentioned earlier, bottlenecks represent the busiest elements in a system. Consequently, a bottleneck subsystem or workstation tends to accumulate a large pile of WIP units upstream and has few or no WIP units downstream. Similar patterns can be observed in throughput rates. If a subsystem is slightly slower than its upstream counterpart, the WIP level at the front of the subsystem accumulates, reaching maximum capacity. Conversely, if a subsystem is slightly faster than its upstream system, the WIP level at the front can be low or even empty. Similar analyses can be applied to downstream buffer WIP levels. Understanding this phenomenon facilitates the rapid identification of bottlenecks in a production flow.

Figure 3.18 shows an example of WIP measurement. During this period, the buffer is frequently full or close to full, indicating that its downstream subsystem is either slower, experiencing more downtime, or a combination of both. Such information proves valuable in locating underperforming elements or bottlenecks.

3.4.1.2 WIP Change and Bottleneck

To simplify the discussion on WIP status changes, consider two subsystems connected by a conveyor, as illustrated in Figure 3.19. Both Subsystems A and B have equal design throughput capacity. Figure 3.20 shows three basic situations (a), (b), and (c) based on the status of WIP units on the conveyor.

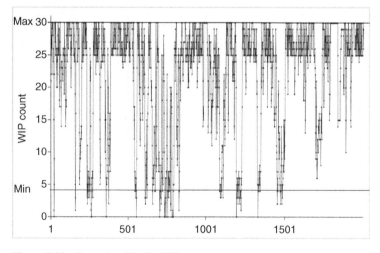

Figure 3.18 Example of buffer WIP count measurement.

Figure 3.19 Two subsystems connected with a conveyor (buffer).

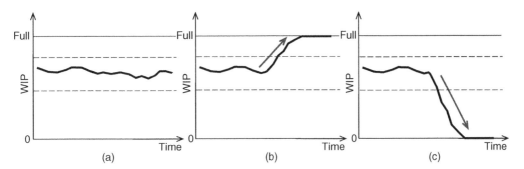

Figure 3.20 WIP status and trends on a conveyor.

- Figure 3.20a: Indicates normal operation, where both Subsystems A and B run smoothly with no issues. The number of WIP units on the conveyor stays at a constant level, with minor fluctuations attributed to random factors.
- Figure 3.20b: The WIP status change suggests that Subsystem B stops or slows down significantly while Subsystem A keeps producing. The number of WIP units on the conveyor noticeably increases as Subsystem B does not take units or takes them at a slower rate.
- Figure 3.20c: Similar to Figure 3.20 situation but reverse. Subsystem A stops or slows down considerably, causing the conveyor to run empty as Subsystem B takes WIP units at a normal rate.

These scenarios can be generalized to various manufacturing systems assuming equal throughput capacity for two neighboring subsystems. The system balanced design related to these situations will be discussed in greater detail in Chapter 8.

The WIP level changes can be further examined for their trends. Here is another example for discussion. Figure 3.21 shows two situations involving three subsystems in a serial setting.

Figure 3.21a indicates the gradually increasing WIP in Buffer 2 and decreasing WIP in Buffer 3. This suggests that Subsystem C becomes slower but is not completely stopped. In this case, Subsystem C is the bottleneck for the entire system. It is also interesting to observe that Buffer 1, although away from Subsystem C, shows a similar trend to Buffer 2 and with a delay.

Following the same reasoning, in the case shown in Figure 3.12b, Subsystem B is the bottleneck. In addition, the sudden changes in WIP levels in Buffers 1 and 2 suggest that Subsystem B might be down. The timestamp of WIP information can help determine when a subsystem's operational status significantly changed.

3.4.1.3 Discussion of WIP Changes

From the example discussion above, WIP status, change, and change rate provide useful information for identifying problematic subsystems, discerning types of problems, and facilitating bottleneck identification. In complex real-world situations, WIP information might not be 100% accurate but can serve as a reliable warning sign, promoting additional investigations. Further assessment, including a comparison with other operations and root cause analysis, is needed to confirm this.

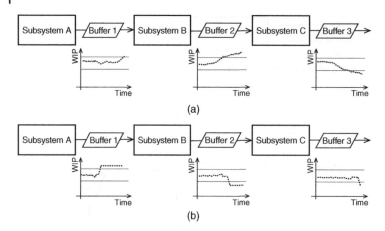

Figure 3.21 Examples of buffer WIP trend monitoring for bottleneck identification.

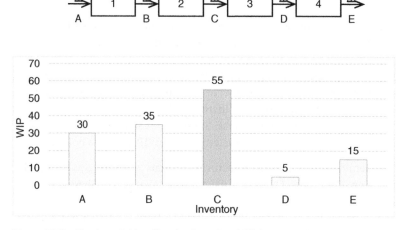

Figure 3.22 Bottleneck identification based on WIP inventory level.

This WIP analysis approach for bottleneck identification is also applicable to non-conveyance inventory and storage. Inventory levels can be measured and tracked in several ways such as the count of WIP units, the number of carrying pallets, or the average WIP waiting time.

Figure 3.22 illustrates an example with four subsystems and five inventory stocks. In this case, Inventory C, located between Subsystems 2 and 3, has the highest level of inventory. It suggests that Subsystem 3 may be the slowest- or lowest-performing subsystem or bottleneck. Subsystem 3 may have the longest cumulative downtime, the slowest CT, a heavy (unbalanced) workload, or a combination of these factors.

It is important to note that the WIP level on a conveyor or in inventory is related to production planning and control, but that is a different concern and falls outside the scope of the book.

Subsection Summary: Monitoring WIP in conveyors is crucial, and analyzing its levels and trends enables effective identification of bottlenecks. The subsequent subsection 3.4.2 delves deeper into buffer WIP status.

3.4.2 Analysis of Buffer Status

3.4.2.1 WIP Change Rate Calculation

The WIP level on a conveyor changes in real time. If a change does not appear random but exhibits a significant trend, this change implies the system's operational status is no longer normal, indicating potential major throughput issues. Thus, addressing the change rate and criteria of WIP levels in a buffer can serve as an informative indicator of a system's working status.

To quantify changes in WIP levels, a method is recommended for calculating the WIP change rate, denoted as R_{WIP}. This rate is defined as a ratio:

$$R_{\mathrm{WIP}} = \frac{\Delta \mathrm{WIP}}{\Delta t}$$

Here, $\Delta \mathrm{WIP}$ represents the number of WIP changes during Δt, which is a predefined time interval. The rate can be visually illustrated in Figure 3.23. The specific interval Δt can be determined based on the production volume and the size of the conveyor. For example, in a production system with a TR of 72 JPH, an initial time interval of three or four minutes can be used. It can be adjusted later, as necessary.

For instance, if $\Delta \mathrm{WIP} = 7$ units and $\Delta t = 3$ minutes, then the WIP change rate is calculated as:

$$R_{\mathrm{WIP}} = \frac{\Delta \mathrm{WIP}}{\Delta t} = \frac{7}{3} = 2.33 \left(\frac{\mathrm{units}}{\mathrm{min}} \right)$$

Using the WIP change rate, it is important to set an alarm threshold within the production monitoring system. When a WIP change rate exceeds the threshold, the monitoring system triggers alerts for the attention of operational managers. Obviously, the threshold value is specific to each case and determined based on a particular operation.

The appropriate alarm threshold may be determined based on experience and through a trial-and-error process. For instance, a starting point for the alarm threshold might be set at two-thirds of the throughput rate. The consideration is that if either the upstream or downstream system stops, the WIP change rate would be equal to the throughput rate, triggering the two-third threshold.

For instance, a system operates at a TR of 72 JPH (or 1.2 jobs/min). An initial threshold for a downtime alarm may be set at 2.4 units/min. The threshold can be fine-tuned to achieve the desired

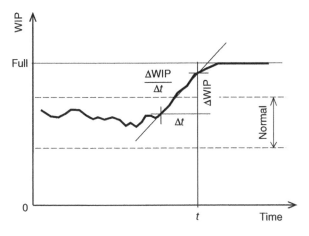

Figure 3.23 Buffer WIP change rate as an operational status monitoring indicator.

sensitivity to downtime. In addition, a separate warning threshold may be established to monitor slowdown issues.

3.4.2.2 WIP Distribution Analysis

In addition to analyzing WIP changes, examining the WIP distribution in a buffer can also provide insights into system performance status. An effective approach is a histogram analysis of WIP distribution on a buffer.

Constructing a histogram to analyze WIP distribution is a straightforward process that can be easily done using MS Excel or other software, such as Minitab. Figure 3.24 shows how to create a histogram using Excel in the following four steps:

1. Collect the WIP data of a conveyor.
2. Select "Data Analysis" from the "Data" tab. If the function is not loaded, it can be activated through "File" → "Options" → "Add-ins."
3. Create a bar graphic (Histogram) by filling in the "Input Range" and "Bin Range." For instance,
   o Input Range: the original WIP data, measured in a period
   o Bin Range: the size of the bin, e.g., 0, 1, 2, ... , and buffer capacity
4. Review and draw conclusions based on the resulting histogram.

Figure 3.25 shows an example of a WIP distribution histogram, with a buffer capacity of 30. The WIP levels at 28, 29, and 30 are considered full or nearly full, presenting 29.4% of the observed time.

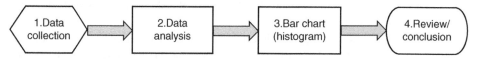

Figure 3.24 Histogram analysis process of buffer WIP using Excel.

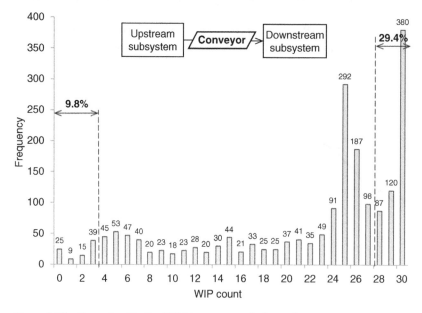

Figure 3.25 Example of buffer WIP histogram analysis result.

Using the formula (Min $= \frac{t_{transfer}}{CT} + 1$) discussed in subsection 3.3.2, the minimum buffer requirement is three WIP units. Accordingly, the buffer is ineffective as empty or nearly empty when the WIP units are below or equal to three, which occurs for 9.8% of the period.

In this example, the WIP distribution does not follow a normal distribution, frequently having too many WIP units in the buffer. The upstream subsystem is blocked for 29.4% of the time as the buffer is almost full. On the contrary, the downstream subsystem is starved for 9.8% of the time as the buffer is almost empty. Overall, this suggests that the downstream subsystem is slower and/or experiences more downtime and should be addressed first. Applying the same analysis on multiple conveyors in the entire system, the bottleneck subsystem can be located.

3.4.2.3 Discussion of WIP Distribution

As mentioned before, the WIP distribution in the buffers ideally exhibits a normal distribution over time. It is under these conditions that the upstream and downstream subsystems are designed with the same throughput capacity and maintained well. Otherwise, the WIP distribution curve may be skewed to one side or exhibit complex shapes. Figure 3.26 illustrates the three basic situations.

Figure 3.26b and c show skewed distributions of WIP, indicating that one of the two connected subsystems has lower throughput performance. In the cases of multiple throughput incidents throughout a period, the distribution pattern can become more complex with multiple peaks and valleys. In such cases, either narrowing down the period selected or excluding known issues can have clearer analysis results.

As previously discussed, a smooth operation should have a stable WIP level in conveyors. Accordingly, a WIP range for normal operations may be set. The normal WIP range may be defined as $\pm 2\sigma$ of the average WIP, covering over 95% of the time analyzed. The σ presents the standard deviation of WIP data in the buffer over an extended period. Due to the connected subsystems not well balanced by system design and/or production planning, the left of the normal WIP range is not necessarily at the midpoint between the Min and Max.

If the WIP level exceeds the established range, it is a warning sign that the system's operation is off normal operations. Figure 3.27 depicts an example of the normal WIP range.

It is important to iterate that establishing a normal WIP range assumes the distribution follows a normal pattern. If the WIP distribution deviates significantly from normal, it may be difficult to set a range until the underlying issues causing abnormal throughput are resolved.

Subsection Summary: Quantitative WIP analysis methods, including analyzing WIP change rates and constructing distribution histograms, deepen understanding of the operational status and enhance the effectiveness of bottleneck identification.

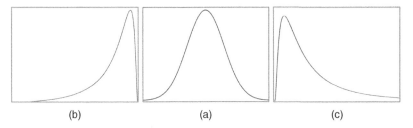

(b)   (a)   (c)

Figure 3.26 Possible WIP distributions in buffer.

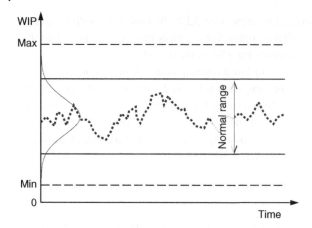

Figure 3.27 Range of WIP units in the buffer during normal operations.

Chapter Summary

3.1. Understanding of Bottleneck

1. Bottlenecks constrain system performance; identifying and improving them enhances overall performance.
2. The TOC is a methodology that involves five steps to identify, exploit, and improve constraints or bottlenecks to enhance system performance.
3. Bottlenecks can be related to throughput, quality, cost, productivity, and/or utilization. They can be caused by equipment, processes, technology, and/or management.
4. Two types: performance bottlenecks (operations related) and capacity bottlenecks (design related). Cross-functional collaboration is needed to address them.
5. Studying the duration and frequency of a bottleneck can help identify its main factors. Momentary and persistent bottlenecks require different considerations and solutions.

3.2. Bottleneck Identification

6. Identifying bottlenecks in large, complex systems is challenging due to dynamic interactions between subsystems and location changes of bottlenecks.
7. Methods like the active period analysis and turning point method can help identify bottlenecks.
8. Multiple KPIs can be analyzed collectively to identify bottlenecks from different perspectives.
9. OEE and standalone OEE are comprehensive KPIs recommended for bottleneck identification and throughput improvement.
10. Bottlenecks and their impacts can vary in time and location, addressing multiple top bottlenecks at the same time may be effective.
11. Computer simulation and mathematical modeling are popular research methods to predict bottlenecks, although they have some limitations.

3.3. Understanding of System Buffer

12. Conveyors serve dual functions – transferring WIP and buffering between subsystems against performance fluctuations.

13. Buffers can take the form of WIP units, inventory, excess process capacity, extra time, or a combination of these.
14. Buffer size, determined by conveyor capacity, affects the extent of the buffering effect on reducing starved/blocked times of connected subsystems.
15. The minimum number of WIP units in a buffer can be estimated based on factors such as buffer size and production rate.
16. Quantitative analysis is possible to evaluate buffering effects, but simulations and experiments are necessary for complex real-world systems.

3.4. Buffer Analysis for Bottleneck Identification

17. Monitoring WIP levels and trends on buffers provides valuable insights into operational status and locating bottlenecks. Gradual or sudden WIP accumulations indicate downstream bottlenecks.
18. Analyzing WIP distribution histograms reveals the percentage of time a conveyor is full or empty. This indicates the relative throughput performance of connected subsystems.
19. Quantifying buffer WIP changes over time into a rate, and setting a threshold, effectively signals abnormal operations like downtime. Useful for monitoring.
20. A normal WIP range can be defined during stable operations. Exceeding this range signals abnormal operations.

Exercise and Review

Exercises

3.1 A system comprises five workstations, each with cycle times of 52, 55, 58, 50, and 51 seconds, respectively. Estimate the throughput rates of the system and identify which workstation serves as the throughput bottleneck.

3.2 Four manufacturing subsystems, arranged in series, have active periods accounting for 95%, 89%, 91%, and 85% of the production time in a week, respectively. Determine which subsystem acts as the throughput bottleneck?

3.3 A system consists of seven operations. Their starved and blocked times (due to various reasons) over a week of production are listed in Table 3.3. Using the concept of a "turning point," identify the operation that is the bottleneck and provide a rationale for your choice.

3.4 A conveyor, which has a transfer time of one minute between two systems with a cycle time of 55 seconds, is in operation. What would be the recommended minimum quantity of WIP units to support system throughput?

Table 3.3   For Exercise 3.3

Operation	1	2	3	4	5	6	7
Starved time (minutes)	50	70	20	40	30	40	50
Blocked time (minutes)	40	30	50	70	60	60	50

3.5   A system that includes four subsystems arranged in series, operates in a continuous flow. The five conveyors associated with these subsystems hold WIP units of 40, 65, 10, 20, and 32, respectively (refer to Figure 3.22). Identify the bottleneck in the system's throughput.

3.6   A conveyor, with a capacity equivalent to 15 minutes of production time and typically filled to two-thirds of its capacity, is in operation. Determine the duration for which the conveyor can compensate for the downtime of its upstream and downstream systems.

3.7   A production line functions at a rate of 50 units per minute, and the WIP change alarm is set to activate at three-fourths of the production rate. If the number of WIP units on the conveyor drops by 120 units within three minutes, would the WIP change alarm be triggered?

Review Questions

(The chapter covers these topics. For further discussion, it is recommended to seek additional information and examples. Diverse perspectives are encouraged.)

3.1   Explain the principle of the Theory of Constraints and provide an application example.

3.2   Describe the types of bottlenecks in a system and discuss which are more important.

3.3   Differentiate between performance bottlenecks and capacity bottlenecks, and provide examples based on manufacturing operations.

3.4   Explain monetary and persistent bottlenecks, their characteristics, and main influencing factors, and provide examples.

3.5   Provide an example of how a change in the quantity of WIP can be used to identify a monetary bottleneck.

3.6   Discuss the relationship between cycle time and throughput rate in a system, providing an example.

3.7   Examine how bottlenecks can be identified based on the active time of operations.

3.8   Explain how to identify bottlenecks based on blockage and starvation information.

3.9   Review the identification of bottlenecks associated with multiple KPIs, providing an example.

3.10   Explain why the identification of bottlenecks can differ when using OEE and standalone OEE with an example.

3.11   Describe how the use of buffers can mitigate the impact of starvation and blockage on a system.

3.12 Explain the minimum WIP requirement for a buffer (conveyor).

3.13 Review how the maximum number of WIP units in a buffer can influence throughput performance.

3.14 Discuss how the number of WIP units in a buffer influences its cushioning effect.

3.15 Explain how the placement of a buffer, whether at the end, the front of a system, or both, can enhance the system's throughput performance.

3.16 Explain how to identify a bottleneck in a system with varying buffer levels, as shown in Figure 3.22.

3.17 Review the potential application of the buffer WIP change rate for the quick detection of a bottleneck.

3.18 Examine the implications of a WIP distribution that is not normal, as determined from a histogram analysis.

3.19 Explain how a skewed distribution of WIP on a conveyor can be utilized for bottleneck identification.

3.20 Discuss the feasibility and significance of establishing a normal range and change rate criterion for WIP units in a buffer as a warning sign (refer to Figure 3.27).

References

Alden, J.M., Burns, L.D., Costy, T., Hutton, R.D., Jackson, C.A., Kim, D.S., Kohls, K.A., Owen, J.H., Turnquist, M.A. and Veen, D.J.V. 2006. General motors increases its production throughput. Interfaces (Providence), 36(1), pp. 6–25. https://doi.org/10.1287/inte.1050.0181.

Betterton, C.E. and Silver S.J. 2012. Detecting bottlenecks in serial production lines – a focus on interdeparture time variance. International Journal of Production Research, 50(15), pp. 4158–4174. https://doi.org/10.1080/00207543.2011.596847.

Bouasria, M., Arvind, A. and Zaka, R. 2020. Bottlenecks to financial development, financial inclusion, and microfinance: a case study of Mauritania. Journal of Risk and Financial Management 13(10), p. 239. https://doi.org/10.3390/jrfm13100239.

Eskandari, H., Babolmorad, N. and Farrokhnia, N. 2013. Bottleneck analysis in a pharmaceutical production line using a simulation approach. Simulation Series, 45, pp. 195–202.

Goldratt, E.M. and Cox J. 1992. The Goal: A Process of Ongoing Improvement. Revised ed. Great Barrington, MA: North River Press.

Gutenberg, E. 1951. Grundlagen der Betriebswirtschaftslehre. Erster Band: Die Produktion (Basics of Business Administration. Volume One: The Production), 1st ed. Berlin: Springer-Verlag.

Hilmola, O.P. and Gupta, M. 2015. Throughput accounting and performance of a manufacturing company under stochastic demand and scrap rates. Expert Systems with Applications, 42(22), pp. 8423–8431.

Hinckeldeyn, J., Dekkers, R., Altfeld, N. and Kreutzfeldt, J. 2014. Expanding bottleneck management from manufacturing to product design and engineering processes. Computers and Industrial Engineering, 76, pp. 415–428. https://doi.org/10.1016/j.cie.2013.08.021.

Kelley Jr, J.E. and Walker, M.R. 1959. Critical-path planning and scheduling. In Papers presented at the December 1–3, 1959, Eastern joint IRE-AIEE-ACM Computer Conference, pp. 160–173. Boston MA. 1–3 December 1959.

Kohls, K. 2020. GM's Throughput Improvement Process. https://www.linkedin.com/pulse/gms-throughput-improvement-process-kevin-kohls/. Accessed June 2022.

Kuo, T.C., Hsu, N.Y., Li, T.Y. and Chao, C.J. 2021. Industry 4.0 enabling manufacturing competitiveness: delivery performance improvement based on the theory of constraints. Journal of Manufacturing Systems, 60, pp. 152–161.

Lawrence S.R. and Buss A.H. 1994. Shifting production bottlenecks: causes, cures, and conundrums. Production and Operations Management, 3(1), pp. 21–37. https://doi.org/10.1111/j.1937-5956.1994.tb00107.x.

Li L, Chang Q, Xiao G and Ambani S. 2011. Throughput bottleneck prediction of manufacturing systems using time series analysis. Journal of Manufacturing Science and Engineering, 133. https://doi.org/10.1115/1.4003786.

Li, L., Chang, Q. and Ni, J. 2009. Data driven bottleneck detection of manufacturing systems. International Journal of Production Research, 47(18), pp. 5019–5036. https://doi.org/10.1080/00207540701881860.

Mehrbod, S.-F. and Tory, M. 2020. BIM-based building design coordination: processes, bottlenecks, and considerations. Canadian Journal of Civil Engineering = Revue Canadienne de Génie Civil, 47(1), pp. 25–36. https://doi.org/10.1139/cjce-2018-0287.

Merriam-Webster n.d. https://www.merriam-webster.com/dictionary/buffer. Accessed June 2022.

Mourtzis, D. 2020. Simulation in the design and operation of manufacturing systems: state of the art and new trends. International Journal of Production Research, 58(7), pp. 1927–1949.

Mütze, A., Lebbing, S., Hillnhagen, S., Schmidt, M. and Nyhuis, P. 2023. Modeling interactions and dependencies in production planning and control: an approach for a systematic description. In Smart, Sustainable Manufacturing in an Ever-Changing World: Proceedings of International Conference on Competitive Manufacturing (COMA'22), pp. 31–44. Cham: Springer International Publishing.

Myrelid, A. and Olhager, J. 2019. Hybrid manufacturing accounting in mixed process environments: a methodology and a case study. International Journal of Production Economics, 210, pp.137–144.

Naranje, V. and Sarkar, S. 2019. Applying Theory of Constraints to a Complex Manufacturing Process. In: Digital Manufacturing and Assembly Systems in Industry 4.0, pp. 75–100, CRC Press.

Padró, R., Tello, E., Marco, I., Olarieta, J. R., Grasa, M. M. and Font, C. 2020. Modelling the scaling up of sustainable farming into agroecology territories: potentials and bottlenecks at the landscape level in a Mediterranean case study. Journal of Cleaner Production, 275, p. 124043.

Plomp, B.I. 2019. A bottleneck analysis to increase throughput at Apollo Vredestein BV (Master's thesis, University of Twente). http://essay.utwente.nl/79836/. Accessed April 2023.

Rogg, J. G., Huckman, R., Lev, M., Raja, A., Chang, Y. and White, B. A. 2017. Describing wait time bottlenecks for ED patients undergoing head CT. The American Journal of Emergency Medicine, 35(10), pp. 1510–1513.

Roser, C., Lorentzen, K., Lenze, D., et al. 2017. Bottleneck prediction using the active period method in combination with buffer inventories. In: Lödding, H., Riedel, R., Thoben, KD., von Cieminski, G., Kiritsis, D. (eds) Advances in Production Management Systems. The Path to Intelligent,

Collaborative and Sustainable Manufacturing. APMS 2017. IFIP Advances in Information and Communication Technology, Vol. 514. Springer. https://doi.org/10.1007/978-3-319-66926-7_43

Roser, C., Nakano, M, and Tanaka, M 2001. A practical bottleneck detection method. In Proceedings of the 2001 33rd Winter Simulation Conference, Phoenix, Arlington, VA, USA, pp. 949–953.

Schultheiss, J. and Kreutzfeldt, J. 2009. Performance improvement in production systems through practice-oriented bottleneck management. In Proceedings of the 4th European Conference on Technology Management, Glasgow.

Silva, G.V., Thomitzek, M., Abraham, T. and Herrmann, C. 2021. Bottleneck reduction strategies for energy efficiency in the battery manufacturing. Procedia CIRP, 104, pp. 1017–1022.

Subramaniyan, M., Skoogh, A., Muhammad, A.S., Bokrantz, J. and Bekar, E.T. 2019. A prognostic algorithm to prescribe improvement measures on throughput bottlenecks. Journal of Manufacturing Systems, 53, pp. 271–281.

Subramaniyan, M., Skoogh, A., Salomonsson, H., Bangalore, P. and Bokrantz, J. 2018. A data-driven algorithm to predict throughput bottlenecks in a production system based on active periods of the machines. Computers and Industrial Engineering, 125, pp. 533–544.

Tang, H. 2018. Manufacturing System and Process Development for Vehicle Assembly, ISBN: 978-0-7680-8346-0, Warrendale, PA: SAE International

Urban, W. 2019. TOC implementation in a medium-scale manufacturing system with diverse product rooting. Production & Manufacturing Research, 7(1), pp. 178–194.

Wang, J.Q., Song, Y.L., Cui, P.H. and Li, Y. 2021. A data-driven method for performance analysis and improvement in production systems with quality inspection. Journal of Intelligent Manufacturing, pp. 1–15.

Wedel, M., Hacht, M.V., Hieber, R., Metternich, J. and Abele, E. 2015. Real-time Bottleneck detection and prediction to prioritize fault repair in interlinked production lines. Procedia CIRP, 37, pp. 140–145.

4

Quality Management and Throughput

4.1 Quality Management

4.1.1 Understanding of Quality

4.1.1.1 Common Definitions of Quality

Quality can be defined in various ways, from different perspectives. The American Society for Quality (ASQ) defines quality as "a subjective term for which each person or sector has its own definition. In technical usage, quality has two meanings: (1) the characteristics of a product or service that bear on its ability to satisfy stated or implied needs; (2) a product or service free of deficiencies." [ASQ n.d.]. Similarly, the international standard ISO 9000:2015 states that "The quality of an organization's products and services is determined by the ability to satisfy customers and the intended and unintended impact on relevant interest parties." [ISO 2015a]. From another perspective, quality can also be assessed in terms of its integration into the ecosystem and societal values [Martin et al. 2020].

At its core, quality should:

- Be customer satisfaction-oriented
- Adhere to product or service specifications
- Reflect various degrees of excellence
- Have multiple dimensions

When examining quality in the context of manufacturing, it can be defined as the characteristics that enable a product to meet customer needs and design specifications, reflecting the first two bullet points mentioned above. In addition, quality is a key enabler of operational excellence and manufacturing throughput (the third bullet point). Addressing the individual aspects and dimensions of product quality must be integrated into specific manufacturing processes.

4.1.1.2 Dimensions of Quality

Because product quality can be viewed and addressed from various aspects and angles, it has multiple dimensions. For instance, Table 4.1 lists eight quality dimensions of a passenger vehicle [Tang 2017].

Quality dimensions are case-dependent. In a manufacturing operation, for instance, the quality attributes of a machined part may include dimensional accuracy concerning shapes and sizes, surface roughness or finish, and hardness.

Chapter 2 discussed an example of building the relationship between door quality and car assembly quality, as demonstrated in Figures 2.20 and 2.21. Other studies on the subject include visual

Manufacturing System Throughput Excellence: Analysis, Improvement, and Design, First Edition. Herman Tang.
© 2024 John Wiley & Sons, Inc. Published 2024 by John Wiley & Sons, Inc.
Companion website: www.wiley.com/go/Tang/ManufacturingSystem

Table 4.1 Key dimensions of product quality with examples (Tang, 2017/with permission of SAE International).

Dimension	Description	Example (car quality)
Performance	Primary operating characteristics	Acceleration: 0–60 mph (0–100 km/h) in 6.8 seconds
Safety	Crashworthiness and crash avoidance (performance)	Safety rating of crash tests by NHTSA and IIHS
Features	Secondary performance characteristics	Folding seats and DVD/TV/Bluetooth function
Reliability	Reliability in working without major failure	Running three years without a major issue
Durability	Measure of vehicle life (replacement preferred over repair)	Engine 95% reliable (without major issues) in 3 year
Aesthetics	Based on appearance, feeling, sound, etc.	Flaming red color (subjective)
Conformance	Meet established standards and expectations	No water leaks
Serviceability	All related to services, including cost, speed, service professionalism	Routine service from a dealer

analytics in the reverse engineering process [Ruediger et al. 2021], uncertainty analysis [Li et al. 2019], and finite element analysis (FEA) on structural quality [Franz and Wartzack 2022].

Manufacturing involves the realization of products and the assurance of their quality through various processes. Throughout manufacturing processes, each dimension of product quality must be controlled through specific processes. Consequently, it becomes essential to establish a connection between individual dimensions of product quality and the corresponding manufacturing process parameters. This integrated approach ensures that quality is systematically integrated into the product throughout its production.

4.1.1.3 Quality and Operational Excellence

Quality itself plays a pivotal role in operational excellence. Numerous studies have shown that quality efforts directly impact profit growth, productivity, and efficiency. A joint global survey conducted by Forbes and ASQ, which included 1869 executives and quality professionals, showed that quality has a direct effect on profit growth. Organizations embracing continuous improvement (CI) or performance excellence practices are more likely to achieve higher levels of productivity [Forbes Insights 2017]. Another survey, based on 324 responses, revealed a positive effect of total quality management (TQM) practices on organizational excellence [Samawi et al. 2018]. According to Fawzy and Olson [2018], Apple's success has been attributed to its emphasis on quality, incorporating the principles of quality management, even if not strictly adhering to Dr. Deming's steps in textbook fashion.

Moreover, quality significantly impacts various performance aspects. Specifically for manufacturing, quality management and adherence to good quality help optimize production by the following:

- Reducing waste from defects and scrap
- Minimizing production disruptions caused by quality issues

- Enabling process stability and statistical control
- Allowing full process capacity utilization
- Maximizing asset utilization through OEE
- Freeing up resources for CI initiatives

The positive link between quality and throughput excellence has been long established [Shingo 1986], as the saying goes: "Quality and productivity: two sides of the same coin" [Butler-Bowdon 2017]. However, the precise quantitative relationships between quality and throughput excellence require a case-by-case analysis based on system settings and process characteristics. A comprehensive assessment is necessary to optimize quality investments for maximum throughput performance.

4.1.1.4 Manufacturing Quality
From a manufacturing perspective, product quality consists of the following three elements:

- First-time quality (good product units without repair or rework)
- Rework/repair quality (good units after rework)
- Scrap (bad units not worth reworking)

Figure 4.1 illustrates their relationship. Further discussion on first-time quality and rework processes is provided in subsection 4.3.2.1.

On the production floor, the primary activities for quality management are to increase good quality units, by inspections, quality control (QC), repair, and CI. These efforts play the following distinct roles:

- Quality inspection evaluates product and part quality through measurement, inspection, and testing.
- Quality control in manufacturing processes maintains consistent adherence to specifications.
- Repair entails reworking or fixing defective units to meet specifications.
- Continuous improvement involves ongoing incremental enhancements to products, processes, and services.

To understand the factors that influence the quality of a manufacturing operation, such as suppliers, processes, and customers, a suppliers-inputs-process-outputs-customers (SIPOC) analysis should be conducted. A SIPOC analysis clarifies the roles of different parties and serves as a foundation for QC and CI. Table 4.2 presents an example of a SIPOC analysis for vehicle manufacturing [Tang 2022]. The scope of a SIPOC analysis can vary, from an operation to an entire plant.

Subsection Summary: Quality, with its diverse definitions and multifaceted nature, is related to and influenced by various processes and aspects of operational excellence, particularly within the manufacturing industry.

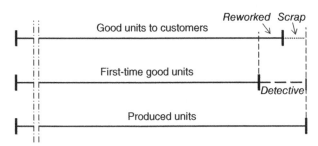

Figure 4.1 Quality categories of product units produced.

Table 4.2 SIPOC of vehicle assembly manufacturing system.

Operation	Supplier	Input	Process	Output	Customer
Part supplier	Part/material manufacturers	Raw materials, parts, etc.	Various	Components, parts, etc.	Vehicle assembly plants
Body shop	Part suppliers	Components, parts, materials	Joining, etc.	Framed car bodies	Paint shop
Paint shop	Body shop, material suppliers	Framed bodies, materials	Painting, sealing, etc.	Painted car bodies	General assembly shop
General assembly shop	Paint shop, part suppliers	Painted bodies and components	Installation, etc.	Completed vehicles	Car buyers

4.1.2 Quality Management System

A quality management system (QMS) is a crucial mechanism for ensuring and maintaining effective manufacturing processes to produce good-quality products. Manufacturing quality can be defined as the conformance of products produced to design specifications based on customer requirements and with technical innovations. Various activities and processes of QMS are in place to ensure the quality of manufacturing.

4.1.2.1 Key Elements of QMS

There are different perspectives regarding quality management in manufacturing systems. ISO 9001 [ISO 2015b] outlines the key elements and requirements for implementing QMS, focusing on processes and documentation. The key elements include the following:

1. Context: Understand internal and external factors affecting QMS results.
2. Leadership: Establish policies, objectives, and processes aligned with the organization's strategic direction.
3. Planning: Identify risks and opportunities, set quality objectives, and create plans to achieve them.
4. Support: Provide resources, competence, awareness, and communications for QMS.
5. Operation: Plan, implement, and manage processes for provision, including design, procurement, production, and release.
6. Performance evaluation: Monitor, measure, evaluate, review, and gather customer feedback through audits, reviews, and more.
7. Improvement: Continuously enhance QMS performance by addressing nonconformities and implementing corrective actions.

Quality pioneer Dr. Joseph Juran introduced the Juran trilogy, consisting of the following three main pillars for a QMS: quality planning, QC, and CI [Juran 1986], as illustrated in Figure 4.2.

In manufacturing, the three pillars play distinctive roles: quality planning is primarily conducted during product design, manufacturing system design, and process planning. QC and CI activities take place on the production floor. Figure 4.3 illustrates a simple example of how these three pillars are integrated into a manufacturing operation.

Manufacturing systems and QMS are often viewed as separate systems. When dealing with manufacturing quality, it is crucial to view QMS as an integral part of the entire manufacturing system

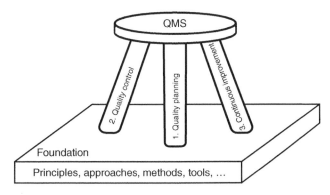

Figure 4.2   Main pillars and foundation of QMS.

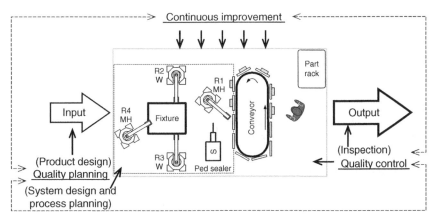

Figure 4.3   Roles of QMS in manufacturing operations.

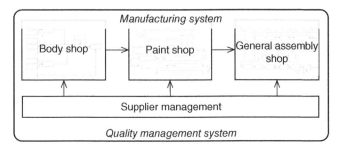

Figure 4.4   QMS and vehicle assembly manufacturing system.

to ensure effective quality assurance. Figure 4.4 shows a high-level view of QMS, covering and supporting the entire vehicle assembly manufacturing systems, including the supplier management system.

4.1.2.2   Discussion of QMS Pillars
The three QMS pillars are not only integral to a QMS but also interact with each other. In general, quality planning is essential to the success of QC and CI. While QC and CI can support each other

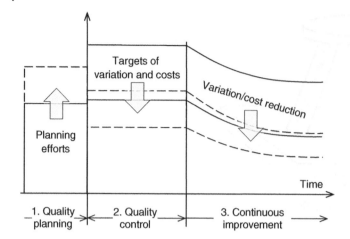

Figure 4.5 Efforts and benefits of improved quality planning.

and realize the goals set in quality planning. From this perspective, quality professionals should have knowledge in all three areas when dedicating one area to be effective.

1. A quality journey begins in the planning stage of product design, establishing quality goals and a roadmap aligned with customer needs. Quality planning defines product performance and manufacturing capabilities, significantly influencing the work and achievements of the other two pillars. As illustrated in Figure 4.5, excellent quality planning can streamline and enhance QC and CI, resulting in fewer quality issues, smaller variation, and lower costs through purposeful design. Consequently, quality planning plays a proactive role in quality assurance.
2. The second pillar of Juran's QMS is QC. QC has long been a widespread practice in different areas of manufacturing industries, involving various activities such as inspections, monitoring, and process adjustments. QC aims to detect out-of-control conditions in operations, analyze root causes, and implement corrective actions. QC is data driven but is fundamentally reactive problem solving. It can be challenging to QC when dealing with imperfect product designs and/or nonrobust manufacturing processes.
3. When manufacturing processes are under statistical control and quality is consistent, the next phase of a QMS emphasizes on CI. The objective of CI is to identify opportunities for enhancing existing performance, taking proactive efforts compared with QC. CI resembles problem solving, aiming for better performance with a mindset of not satisfying the status quo. Chapter 6 will cover a range of principles, approaches, and tools that support QC and CI.

4.1.2.3 Deming's 14 Points
Dr. Deming, a notable figure in the realm of quality management, introduced 14 points that encompass management practices [Deming 1986]:

1. Establish a constant purpose toward improvement
2. Adopt the new philosophy
3. Cease dependence on inspection for achieving quality
4. Minimize total cost instead of making decisions based solely on price
5. Continuously improve and refine systems
6. Provide on-the-job training

7. Adopt and institute leadership
8. Remove fear from the workplace
9. Break down barriers between departments
10. Do away with slogans, exhortations, and workforce targets
11. Eliminate numerical quotas for employees and numerical goals for management
12. Remove barriers that diminish employees' pride in their work
13. Institute a vigorous program of education and self-improvement
14. Engage everyone in the process of transformation

Deming's points extend beyond quality management and hold significance in various aspects of business management. For instance, point 4, stressing the minimization of total cost, is a fundamental principle central to throughput management, a core theme explored throughout this book. Likewise, point 5, emphasizing CI, is a foundational element in QMS, a vital mindset and practice within throughput management, promoting operational excellence. Further discussion on these topics will be provided in subsequent chapters.

4.1.2.4 Quality Management Standards

Guided by quality management principles, various industries embrace formal standards and guidelines for QMS, adapting them to their specific needs and processes. Examples of quality management standards include:

- ISO 13485:2016 Medical Devices Quality Management Systems – Requirements for Regulatory Purposes
- ISO 22000 Food Safety Management
- ISO/IEC 20000-1:2018 Information technology – Service Management – Part 1: Service Management System Requirements
- ISO/IEC 27001 Information Security Management
- AS 9100D Quality Management Systems – Requirements for Aviation, Space, and Defense Organizations

These standards are crucial in ensuring quality, safety, and regulatory compliance. In practice, these QMS standards should be translated into detailed procedures, instructions, criteria, and more, ensuring that quality is upheld throughout operations.

As an example, IATF 16949 is the QMS guidelines for the automotive industry [IATF 2016]. This standard, titled "Quality Management System Requirements for Automotive Production and Relevant Service Parts Organizations," builds upon the ISO 9001 framework. IATF 16949 comprises ten chapters, outlined in Table 4.3.

All elements of a QMS are connected to production throughput, some are more directly related than others. For instance, Sections 8 (Operation), 9 (Performance Evaluation), and 10 (Improvement) of IATF 16949 are closely related to throughput management. Further in-depth discussions regarding this topic can be found in other chapters of this book, in addition to the sections directly addressing quality-related subjects in this chapter.

Subsection Summary: Examining QMS can be conducted through multiple lenses such as key elements, three pillars, Deming's 14 Points, and industry standards. A good understanding of QMS lays a solid foundation for its roles in overall operations, in terms of cost, throughput, and improvement.

Table 4.3 Main contents of IATF 16949:2016.

Chapter (Section)	Title	Content
1–3	Introductions	Scope, Normative References, Terms, and Definitions
4	Context of Organization	Definitions of the requirements for determining the scope and general QMS requirements
5	Leadership	Leadership commitment to a QMS, corporate responsibility, and quality policy
6	Planning	Risks, opportunities, and risk analysis, including requirements for preventive actions, contingency plans, objectives, and plans
7	Support	Requirements for people, infrastructure, work environment, resources, knowledge, competence, and communication
8	Operation	Requirements on planning, product review, design, purchasing, and controlling for monitor and measure a product or service
9	Performance Evaluation	Assessment of customer satisfaction, internal audits, monitoring of products and processes, and management reviews
10	Improvement	Requirements for problem solving, corrective actions, error-proofing, and continual improvement

4.2 Cost Analysis of Quality

Operational management typically emphasizes the efficiency and costs of operations, while manufacturing practitioners use technical terminology and KPIs. Cost analysis is integral to both QMS and overall business operations, involving the translation of quality-related activities into financial benefits, as depicted in Figure 4.6. The translation can readily be understood and communicated within the organization, bridging the gap between these two aspects of the business.

More importantly, finance aligns the quality management department with broader organizational goals. Savings resulting from quality improvements and throughput enhancements bolster the company's profits and pre-tax earnings. Conversely, quality issues adversely affecting system throughput have a negative impact on overall business costs. The combined benefits of reduced costs and increased profitability lead to long-term success.

4.2.1 Categories of Quality Costs

4.2.1.1 Costs of Good Quality and Poor Quality

Cost of quality (COQ) comprises two main categories: the cost for achieving good quality and the cost due to poor quality. The first category represents investments in quality assurance.

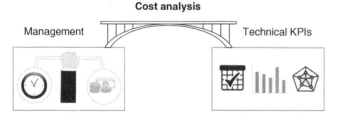

Figure 4.6 Management focuses, cost analysis, and technical KPIs.

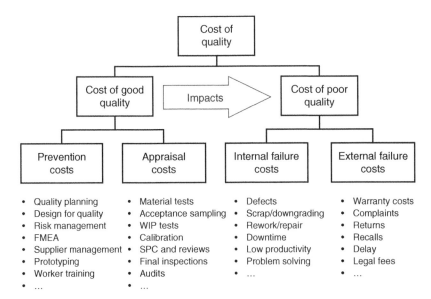

Figure 4.7 Breakdown of quality cost categorization.

The second category measures the monetary losses associated with defects and their impacts on system performance.

These COQ categories can be further divided into four types, as illustrated in Figure 4.7, along with a list of exemplary items. Each item can be represented by or converted into a monetary value as a basis for further cost analysis. This COQ categorization is known as the prevention-appraisal-failure (PAF) model.

The cost of achieving good quality, or "good quality cost," includes the investments to prevent potential quality issues and appraise the quality of products and services. Prevention costs are incurred before an operation, while appraisal costs are incurred during or at the end of an operation. In some cases, appraisal costs may complement prevention costs with both aiming to ensure products conform to specifications.

Ensuring quality during the design phases can be more cost-effective, aligning with the primary goal of quality planning. One important practice is to incorporate quality by identifying and addressing potential quality issues as early as possible. This can also be achieved by implementing quality measures in an early stage of the manufacturing process planning, ensuring products meet the desired quality requirements within the manufacturing capabilities. For more information on these topics, readers can refer to dedicated sources.

On the contrary, the costs due to poor quality, or "poor quality cost," can occur both before and after product delivery to customers, referred to as internal failure cost and external failure cost, respectively. In general, the external failure cost is much higher than the internal failure cost for the same quality problem. Effective appraisal plans and activities can minimize external failure costs by detecting quality problems before delivery to customers.

4.2.1.2 Direct Costs and Indirect Costs

Quality costs can also be categorized as either direct or indirect, depending on their relationship to manufacturing production. Direct costs are closely linked to production operations, while indirect costs support quality assurance for those operations. However, categorizations and perspectives may vary depending on the context of an organization.

In addition, grouping quality costs into two other categories – up-front acquisition investments and operating expenditures – can provide further insights. Up-front acquisition investments are made during system development while operating expenditures are part of routine work within operational management.

Table 4.4 presents common manufacturing activities with different cost categorizations, including PAF, direct/indirect, and up-front/operating. The table also indicates the relationship between

Table 4.4 Category and relationship of quality costs.

Type		Task	Investment		Operations	
			Up-front	Operating	Direct	Indirect
Good quality	Prevention	Planning	√			√
		Design for quality	√			√
		FMEA	√			√
		Error-proofing	√			√
		Maintenance		√		√
		Incoming material test		√		√
		Training	√	√		√
		…				
	Appraisal	Calibration		√		√
		WIP test		√		√
		Inspection	√	√		√
		SPC		√		√
		Audit		√		√
		Supplier quality	√	√		√
		…				
Poor quality	Internal failure	Defect		√	√	
		Scrap		√	√	
		Rework		√	√	
		Downtime		√	√	
		Low productivity		√	√	
		Problem solving		√		√
		…				
	External failure	Warranty		√		√
		Complaint		√		√
		Return/repair		√		√
		Recall		√		√
		Delay		√		√
		Legal fee		√		√
		…				

different categorizations. Some manufacturing activities in the table may belong to different categories in certain operations. Some activities fall under more than one category, which is discussed in the next subsection 4.2.1.3.

4.2.1.3 Considerations in Cost Categories

The categorization of quality costs in Table 4.4 seems straightforward, but it may present challenges in practice. Quality problems may not always manifest as obvious failure costs and can fall into different categories depending on the situation. For example, some problem-solving efforts can be considered both internal failure costs and prevention costs, while certain efforts can fall under both prevention and appraisal categories. As a result, the items included in COQ categories vary based on their applications.

Another challenge pertains to certain indirect quality costs such as those associated with customer satisfaction and company reputation. These may not be easily quantifiable in monetary terms. Estimates, if consistently practiced over time, can serve as valuable indicators to quantify such quality costs.

Internal failure costs and external failure costs are closely related. The costs due to quality issues and defects can be much higher if discovered after shipping. Returned products may be repaired or reprocessed by manufacturers to recover some costs, with a penalty of using additional resources. In supplier quality management, both internal failure costs (at a supplier site) and external failure costs (at their customer site) are often studied and documented in the manufacturing industries.

Furthermore, some quality costs may not always fit neatly into the PAF categories. While the PAF model is widely used, new quality cost types may be introduced in certain situations if the standard categories are not applicable. For instance, one study introduced a "recycle cost" category for the expenses related to material resale, retake, and scrapping [Obied-Allah 2016]. In another study [Rahardjo et al. 2020], authors reclassified the costs as preventive, corrective, and rejection costs using different names.

When considering different investment alternatives in improvement projects, there is often a hidden cost known as "opportunity cost." It can be defined as the difference between the chosen and unchosen investments. A study [Alglawe et al. 2019] showed that if opportunity costs are considered, it may be more beneficial to allocate additional resources to prevention and appraisal.

Subsection Summary: Quality costs can be classified in different ways. For instance, good quality versus poor quality, and direct costs versus indirect costs. The practical application of these concepts involves considerations and challenges, more discussion in the following subsections.

4.2.2 Discussion on Quality Costs

4.2.2.1 Total Quality Cost

In manufacturing, the total quality cost combines both good-quality costs and poor-quality costs:

$$\text{Total COQ} = \text{Good quality costs} + \text{Poor quality costs}$$
$$= (\text{Costs of prevention} + \text{Costs of appraisal})$$
$$+ (\text{Costs of internal failures} + \text{Costs of external failures})$$

Figure 4.8 shows a typical relationship between total COQ and good quality and poor-quality costs in mass production. The figure highlights the lowest point of total COQ, referred to as the economic quality status of a system or shop. Economic quality is often the interest of operational management in reducing quality costs. However, this is unlikely an optimal outcome for the desired high quality.

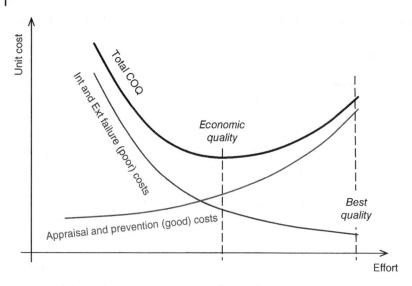

Figure 4.8 Overall quality cost trends and economic quality and best quality.

To achieve the highest possible quality, implying the lowest internal and external quality costs, more appraisal and prevention measures can be necessary. This, in turn, could lead to higher total quality costs, even with low internal and external quality costs. Such a high-quality strategy is particularly applicable to luxury product brands.

Figure 4.8 also indicates an overall inverse relationship between internal and external failure (poor quality) costs and prevention and appraisal (good quality) costs. It is true that higher prevention and appraisal efforts can drive down failure costs. However, the precise cost relationship depends on the specific product design and manufacturing processes involved.

There are numerous studies on cost analysis in manufacturing industries. For example, a study used a closed-loop analysis with four steps in a loop: cost analysis, improvement opportunity, launch project, and project review [Liang et al. 2022]. Another study [Farooq et al. 2017] suggested that the inspection strategy (and associated costs) should be selected based on the quality level of the product. For instance, a double acceptance sampling plan can be justified as an optimal strategy if the external failure costs are remarkably high, and when the range of conformance is between 98.8% and 99.65%. Another study combined costs of quality in the product life cycle toward achieving robust product design, applying the Taguchi design of experiments and the PAF model [Janatyan et al. 2023].

4.2.2.2 Considerations of Cost Relation

Probably not every case agrees with the typical cost relationship presented in Figure 4.8. Higher prevention and appraisal costs do not always result in better quality. For example, Plewa et al. [2016] found that prevention and appraisal cost components were not found to be significantly higher at higher levels of quality. Therefore, it is strongly recommended to study the relationship between good-quality costs and poor-quality costs for a specific operation as a foundation for performance improvement or cost reduction.

Assessing a system's position on a quality cost curve can be a challenging and time-consuming task. Often, conducting such investigations reveals additional issues and raises questions, leading to more tasks and tests that might discourage further investigations. As a result, the company might struggle to accurately assess its quality cost status and relationship to determine an improvement direction.

Furthermore, the interface between bookkeeping accounting and quality work presents gaps in many cases, often related to a short-term, cost-focused mindset. While certain cost items, such as external failure costs, can be precisely recorded, other aspects, such as prevention and appraisal efforts, may not be itemized in an accounting system. This discrepancy can lead to difficulty and confusion in determining the definition of quality costs, acceptable measurements, accurate data interpretation, and evaluations. Consequently, costs of quality and quality enhancement assessment may be viewed as monetary estimates rather than precise accounting dollar figures.

To address these issues, initiatives have been undertaken to better integrate quality systems with accounting systems. A recent study showed the implementation of a system for recording and analyzing quality costs using spreadsheets, formulas, macros, and visual alert/action indicators for a medical device company with over 6000 medical tool designs [Herzog and Grabowska 2021]. The system notified responsible personnel upon detecting adverse trends, thereby aiding proactive quality management.

Subsection Summary: The relationship between good-quality costs and poor-quality costs can be complex, often with a reverse trend between them. The total quality cost reflects QMS strategy and practice. Tailored approaches and financial analysis can benefit quality and overall system performance.

4.2.3 Project Economic Evaluation

When proposing an improvement project, allocate resources to address specific issues. Meanwhile, assess and quantify the benefits of the project, such as increased quality rates, cost savings, or reduced downtime, in monetary terms. It is important to evaluate projects based on their benefits and costs, which will aid in comparing and selecting effective proposals.

4.2.3.1 Break-even Estimation

To assess the financial values of improvement projects, a straightforward method is calculating their break-even points. The break-even point of a project occurs when total benefits equal total costs, resulting in neither gain nor loss at that point. This simple tool is useful for comparing multiple projects.

For an improvement project, its break-even point is calculated as:

$$\text{break-even} = \frac{\text{Total cost}}{\text{Total gains or savings}}$$

For instance, a project requires an initial investment of \$20,000 and is projected to yield monthly savings of \$5000. Applying this formula gives the break-even point for the project:

$$\text{break-even (time)} = \frac{\text{Total cost}}{\text{Total gains or savings}} = \frac{20{,}000}{5000} = 4\,(\text{months})$$

A break-even point can be measured in terms of time (months or years), product units, and monetary terms (e.g., dollars).

The break-even method operates under the implicit assumptions that values remain constant over time and that there are no long-term trends of benefits. However, for projects spanning longer

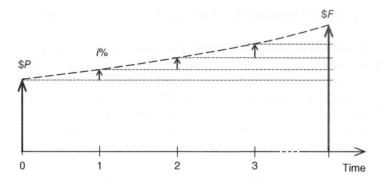

Figure 4.9 Cash flow diagram showing *P*, *F*, and *i*.

periods, such as a couple of years, this method becomes less accurate because the underlying assumptions no longer hold true.

A primary reason for the inaccuracy of the break-even method is its disregard for changes in the value of money over time. A monetary value in the future is not equivalent to its present value. This fundamental financial concept is known as the time value of money.

4.2.3.2 Economic Analysis Basics

To account for changes in the value of money over time, the common approach is to use an interest rate in analysis. An interest rate, expressed as a percentage, represents the rate of change in money's value over time. For example, if an interest rate is 10%, it means that the monetary amount will increase by 10% after a defined period. It is important to note that an interest rate can be compounded over multiple periods.

Figure 4.9 illustrates a simple diagram depicting value changes over time. In this figure, "P" represents the present value, "F" represents the future value, and "i" symbolizes the interest rate. This illustrative representation is usually referred to as a cash flow diagram. The time value of money principle takes interest rates into account, which makes finance analysis more complex than a simple break-even calculation.

Note that improvement projects typically require an initial investment. In such cases, the "P" value is negative, often represented by a downward arrow in a cash flow diagram. Also, the benefit, often presented as "F," from an improvement project may be realized periodically, such as monthly, over the project timeline. This type of economic analysis is a common practice, and valuable for generating accurate predictions.

4.2.3.3 Economic Analysis Application

Here is a simple example for a discussion of economic analysis methods. An improvement project requires an initial investment of $50,000 (*P* = $50k) at the beginning. The company's minimum acceptable rate of return (MARR or required interest rate) is 15%. The projected savings are $25,000 annually (*A* = $25k) for three years. The key question is whether the project is financially justifiable with the MARR.

An economic analysis for this project can be conducted in two ways. The first approach involves calculating the interest rate based on the values of *P* and *A* to see if the actual return rate meets MARR, as depicted in Figure 4.10a. The alternative method is to determine the annual benefit based on the given P and MARR to check the actual benefit, as shown in Figure 4.10b.

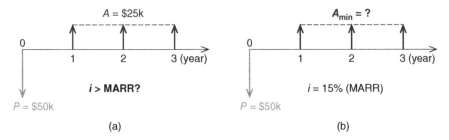

Figure 4.10 Cash flow diagrams of discussion example.

MS Excel can be used for the two corresponding calculations:

1. To calculate the actual return rate, the Excel RATE() function can be used:

   Benefit rate = RATE(3,25, −50) = 23.4%

   This project has a much better rate of return than the required MARR of 15%. Therefore, the project is justified financially.
2. To calculate the annual benefit at a 15% MARR, the Excel PMT() function can be used:

   Annual benefit = PMT(15%, 3, −50) = $21.90k.

The project's annual benefit of $25,000 under the MARR is notably superior to $21,900, confirming its financial justification, as concluded by the other method above.

In real-world scenarios, situations and corresponding calculations can become more complex. For instance, improvement projects may involve investments over multiple periods, with varying benefits, and nonconstant rates over time. However, the fundamental principles of economic analysis remain consistent. Readers can find in-depth information in finance textbooks.

Subsection Summary: Economic analysis can be used for improvement project evaluation, considering benefits, costs, and the time value of money. Break-even analysis is for quick and initial estimates. For greater accuracy, the interest rates should be incorporated into economic analysis.

4.3 Quality in Production Throughput

4.3.1 Quality Contribution to Throughput

4.3.1.1 Quality in System Performance

In an ideal world where product design, process planning, and operational management are flawless without external influences, manufacturing systems should produce high-quality products. However, due to various known and unknown factors and variations, products often encounter quality issues. To mitigate them, manufacturing systems employ two key functions: monitoring and repair in operations. Monitoring functions can be executed through various types of inspections. If a quality defect is identified, it can be repaired if the resources justify such action.

To better understand how quality impacts system throughput, let us examine a practical example. Figure 4.11 shows a manufacturing system with quality inspection and repair functions. This paint system has a capacity of producing 500 units per shift. During a shift, 35 units with defects require light repairs, while 22 other defective units require heavy repairs. In addition, five units are deemed as scrap because of being unrepairable technically and/or economically. In total, 17 heavily repaired units need to re-enter the system for reprocessing.

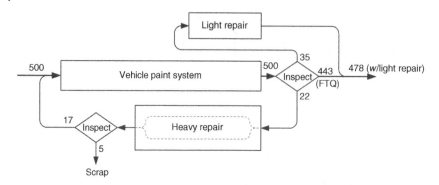

Figure 4.11 Example of quality inspection and repair in manufacturing processes.

In this example, quality affects the system throughput in several ways. First, the loss of five units, equivalent to 1% of system capacity, is due to scrap. Second, the reprocessing of heavily repaired units depletes the system's capability. The utilization of the system in this example is reduced to 96.6% (from 500 to 483). This impact can cascade to subsequent shifts. Third, both light and heavy repair processes require additional resources, thereby increasing the unit cost. This cost increase serves as one of the indicators of the system's quality and throughput performance.

4.3.1.2 Discussion on Quality Impacts

The example in Figure 4.11 simplifies the situation and does not consider several variables such as WIP units in the system, varying repair times, and the dynamic nature of quality issues. For instance, the number of units needing reprocessing varies from shift to shift, affecting subsequent shift operations. To obtain reliable information about the quality's impact on throughput performance, it is advisable to collect data and calculate the mean (average) across multiple shifts.

The quality's impact on system throughput varies depending on factors such as the manufacturing context. In batch processes, the impact can be severe. Quality issues arising from design, process parameters, and/or production control can lead to entire batch failure, which then requires reworking or scrapping. Reworking and reprocessing are much less efficient than standard manufacturing processes. Resources and personnel time devoted to rework subtract from those available for new material production, although it may still be a preferable option compared to scrap.

Scrap presents a more substantial problem, as it wastes raw materials, time, and energy, and generates no output. In addition, disposing of scrap requires additional resources. In such cases, it becomes evident that production would have been significantly better off if the faulty batch had never been produced in the first place.

This discussion emphasizes the paramount importance of initial good quality in a manufacturing system. When a system consistently produces good-quality products, the need for additional resources and costs for repairs can be significantly reduced. Thus, ensuring initial good quality is of utmost importance for optimizing operational efficiency and effectiveness.

4.3.1.3 Quality Impacts on OEE

Quality's impact on system throughput can be assessed via OEE, a direct impact from quality itself, and an indirect impact through the other two elements of OEE. Concerning the direct impact, Table 1.2 in Chapter 1 outlines several quality-related operational states, such as quality faults and quality alerts, which cause production to stop and/or slow down.

In addition, defective WIP units from upstream operations can introduce issues in the current processes, notably affecting loading, positioning, and certain processes, resulting in operational

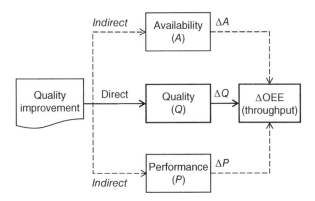

Figure 4.12 Quality contribution pathways to system throughput performance OEE.

slowdowns or downtime. For instance, quality issues in vehicle body shop operations can cause problems in its downstream paint shop and/or general assembly operations, affecting normal processes there.

Moving on to the indirect impacts of quality, quality has sophisticated impacts on system operational availability and speed performance, the other two elements of OEE (Figure 4.12). To attain comprehensive excellence in manufacturing operations, understanding how quality affects system availability and performance is crucial.

Considering both direct and indirect impacts of quality on OEE, a quality improvement involves changes in quality (ΔQ), and associated changes in availability (ΔA) and speed performance (ΔP). As discussed in Chapter 2, the OEE would be: $\Delta OEE \approx \Delta A + \Delta P + \Delta Q$, rather than solely $\Delta OEE \approx \Delta Q$.

A further discussion can delve into the relationship between quality and other KPIs. For instance, increasing the operating speed of a production line seems like a way to increase its throughput rate. However, a higher operating speed can lead to some quality issues, resulting in more defective products that require rework or repair. The overall throughput at the higher operating speed may be lower than that at the normal speed due to the increase in quality issues. Therefore, caution is warranted regarding potential "side effects" on quality or other KPIs when adjusting a process parameter.

Quantifying the indirect impact on system throughput performance resulting from quality is a key aspect of QMS and COQ, remaining challenging. This challenge arises due to the different influences that quality issues and improvements can have. Once these indirect impacts are quantified, they should be monetized and incorporated into the total quality cost, such as internal failure cost, or as a separate category for quality indirect cost. This subject holds significant importance for the manufacturing industry, especially for quality professionals.

4.3.1.4 Research Example Review

Research studies provide further insights into the relationship between quality and throughput. Here are a few study examples in different areas:

- Owen and Blumenfeld [2008] conducted a study in automotive manufacturing, revealing quality-speed trade-off relationships that require in-depth analysis based on actual data.
- Moradinaftchali et al. [2016] investigated constraints and tolerance allocation in machining components that impact both quality and assembly productivity. The authors introduced a multi-step approach, including a "basic algorithm" to identify improvement operations and

subsequent algorithms to maximize productivity gains for each component. The approach, tailored to machining parts and assembly, offers valuable references that can be applicable to the study of various manufacturing operations.

- Quality management also positively influences the well-being, attitudes, and satisfaction of employees, especially those who play a crucial role in manufacturing systems such as production workers, technicians, and engineers. This, in turn, results in enhanced productivity and improved throughput performance. In a case study by Putri et al. [2017], the implementation of TQM had a positive impact on employee productivity within a plant. The study assessed productivity using indicators such as willingness to work, the work environment, and work relationships.
- A case study by Schwerha et al. [2020] introduced tools for integrating safety, productivity, and quality metrics, resulting in various benefits, including enhanced communication. Qualitative advantages of integration noted by the authors include reduced ergonomic risks, improved co-worker interactions, and more purposeful work for employees.

Subsection Summary: Quality management plays a pivotal role in manufacturing system excellence. Quality management significantly influences key aspects and metrics such as throughput, OEE, and employee productivity. As manufacturing systems evolve, additional efforts for deeper exploration in the field are strongly recommended.

4.3.2 Quality Analysis of Serial and Parallel Operations

4.3.2.1 Throughput Yield of Serial Systems
Throughput yield (TPY) represents the rate at which defect-free products are produced on the initial attempt, including minor in-line adjustments or built-in touch-ups in the original processes. This concept is commonly known as first-time quality (FTQ) or first-pass yield (FPY).

In the example shown in Figure 4.11, the system's FTQ is 88.6% (443 out of 500 units). After light repair work, the quality rate increases to 95.6% (478 out of 500 units), and considering all repairable products raises the quality rate to 99.0% (495 of 500 units). Notably, units with offline repairs are excluded from the first-time quality calculation.

Maximizing TPY without the need for repair or reprocessing is an ideal goal in manufacturing. TPY is a focus of QC and is influenced by various factors, including the specific context of manufacturing operations.

For a general serial-setting system, the TPY can be analyzed based on the quality failure rate (p) of an operation. In a serial setting with two operations, the second operation receives $(1 - p_1)$ good units, and its own TPY is $(1 - p_2)$. Then, the two-operation system produces $(1 - p_1) \times (1 - p_2)$ good units.

The TPY of a multi-operation system in a serial setting, denoted as rolled or rolling throughput yield (RTY), is calculated as:

$$\text{RTY} = (1 - p_1) \times (1 - p_2) \times \cdots \times (1 - p_n) = \prod_{1}^{n}(1 - p_i)$$

This formula represents the cumulative throughput yield for the entire system, considering the individual quality rates $(p_1, p_2, \ldots, p_n)$ for each operation (refer to Figure 4.13):

Consider a system with five operations in a serial setting, as illustrated in Figure 4.14. The TPY of each operation is known, as indicated in the figure. The RTY of the entire system can be calculated as:

$$\text{RTY} = 99\% \times 98\% \times 97\% \times 98\% \times 96\% = 88.5\%.$$

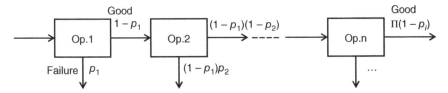

Figure 4.13 Quality rate of a serial system (a Bernoulli process).

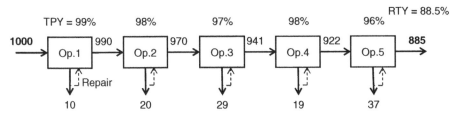

Figure 4.14 Example of a serial manufacturing system TPY and RTY.

The observation pertains to the probability of production yield in a finite sequence of binary exclusive variables "p" and "$1 - p$," known as the Bernoulli process. Similarly, the Poisson process, considering scenarios involving multiple defects, can also be utilized. Interested readers can explore these topics further.

The throughput yield analysis for serial systems is straightforward and serves as a foundation for the same analysis for other configured systems. An effective approach for different configurations is to decouple a system into multiple serial segments, apply the throughput yield analysis on each segment, and then unite the results. Further discussion on this is explored in subsection 7.3.3 of Chapter 7. Parallel systems will be addressed in the next subsection 4.3.2.2.

4.3.2.2 Quality Characteristics of Parallel Systems

Turning now to parallel-setting systems, manufacturing systems often incorporate parallel settings of subsystems for slow-process and/or high-volume productions. In these settings, parallel subsystems are typically identical or remarkably similar in terms of process and functionality for volume production. An example is shown in Figure 4.15, where parallel subsystems A and B produce the same products. The products from the parallel subsystems are mixed and then moved to the next stage.

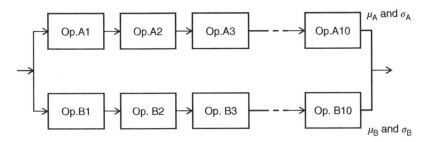

Figure 4.15 Parallel subsystems with individual mean (μ) and standard deviation (σ).

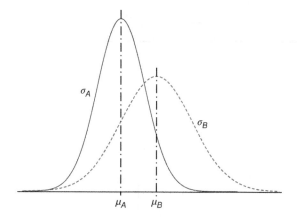

Figure 4.16 Illustration of two different normal distributions.

Unique quality concerns arise in such parallel settings of manufacturing systems. Although parallel subsystems can be designed to be identical, process parameters may differ in individual subsystems on the production floor. The unintentional differences can lead to the parallel subsystems no longer being identical in certain quality attributes. Consequently, parallel subsystems have different means or averages (μ) and variation (in terms of standard deviation σ) for a quality attribute. Figure 4.16 illustrates two different distributions of a quality attribute from two parallel subsystems.

The mean (μ_s) of the entire system, assuming equal usage of subsystems, can be calculated as follows. For a two-parallel system, the calculation is:

$$\mu_s = \frac{\mu_1 \times n_1 + \mu_2 \times n_2}{n_1 + n_2}$$

where n_1 and n_2 are the sample sizes of a quality attribute from the two parallel subsystems, respectively. If the sample sizes are the same, then:

$$\mu_s = \frac{\mu_1 + \mu_2}{2}$$

The variation of parallel systems is complicated. For two parallel subsystems, the system variation (σ_s) of combined products can be estimated using the pooled standard deviation:

$$\sigma_s = \sqrt{\frac{(n_1 - 1) \times \sigma_1^2 + (n_2 - 1) \times \sigma_2^2}{n_1 + n_2 - 2}}$$

where σ_1 and σ_2 are standard deviations of a quality attribute from the two parallel subsystems, respectively. If sample sizes are the same and large (e.g., ≥ 100), then the equation can be simplified as:

$$\sigma_s \approx \sqrt{\frac{\sigma_1^2 + \sigma_2^2}{2}}$$

For example, the means and standard deviations of a quality attribute from two parallel subsystems are known with $n_1 = 55.2$, $n_2 = 57.4$; $\sigma_1 = 5.41$, $\sigma_2 = 4.92$. The system mean would be 56.3, and the system standard deviation would be 5.17, calculated based on the formulas above.

For three or more parallel subsystems, both equations for μ_s and σ_s can be expanded accordingly. Mathematically, the equation for pooled standard deviation is based on the assumptions of full independence of different data sets and equal mean value from the same population. Caution is required when using the equation, as its accuracy relies on these assumptions. The next subsection 4.3.2.3 discusses situations where the mean values from parallel subsystems are unequal.

4.3.2.3 Variation Reduction in Parallel Systems

Building on the previous subsection 4.3.2.2, the challenge of managing variation becomes more critical for QC compared with the mean values of a quality attribute from parallel subsystems. Moreover, the presence of different means can exacerbate the situation, resulting in substantially higher variation. This also leads to reduced analysis accuracy as it deviates from the assumption of the pooled standard deviation.

For simplicity, let us assume that a product quality attribute from two parallel subsystems follows the same distribution with the same variation but different mean values. As illustrated in Figure 4.17, the overall trend shows the combined variation of two parallel subsystems under different levels of the mean ($\Delta\mu$) based on data simulation.

From Figure 4.17, when two means are 2σ apart ($\Delta\mu = 2\sigma$), the combined variation is 40% higher than the baseline ($\Delta\mu = 0$). While the figure provides a basic idea, in practice, data distribution, μ, and σ of each measurement set can vary, resulting in greater and more complex combined variation than shown in this illustration.

Figure 4.18 illustrates an actual example. There are two assembly lines, North and South. Each line has its own measurement data set. The figure shows the variations of individual and combined datasets in terms of 6σ.

In this example, at measurement point 22 U/D (marked A in the figure), the combined variation is significantly larger than the individual variations, suggesting significant differences in the means of two individual measurements. In such cases, efforts to reduce variation should first focus on aligning both means before addressing individual variation.

On the contrary, at point 36 C/C (marked B), the combined variation is close to the individual variations, suggesting that the difference between two individual means is small. Accordingly, directly reducing individual variation may be an effective strategy.

Subsection Summary: Throughput yield is a key quality characteristic for an operation, while rolled throughput yield is for a system. Serial and parallel systems exhibit distinctive characteristics, with parallel systems requiring unique considerations for the combined variation from parallel subsystems to mitigate overall variation.

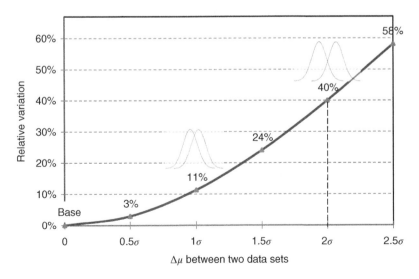

Figure 4.17 Influence of mean differences on overall variation.

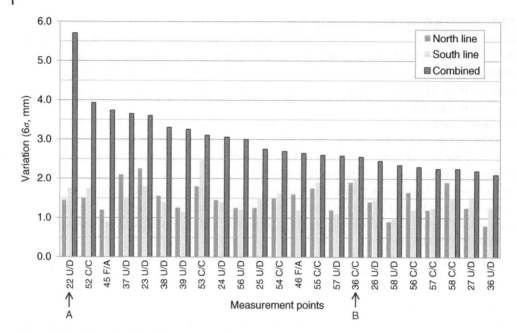

Figure 4.18 Example of individual and combined variations of two systems.

4.4 Discussion of Improving Quality

This section encompasses proactive and reactive quality methods, including inspections, audits, error-proofing, and data analysis tools. These approaches are aimed at achieving CI in quality and throughput, ultimately contributing to operational excellence.

4.4.1 Quality Appraisals

4.4.1.1 Product Quality Inspections

A primary type of quality appraisal activity on the production floor, normally integrated into manufacturing processes through system design, is quality inspections. Inspections involve examining, testing, and analyzing quality attributes of products and reporting their findings. The results of inspections are then compared against specified requirements to determine if the product quality conforms to them.

Inspections are applicable to finished products, WIP units, and incoming parts. An example of in-line inspection is illustrated in Figure 4.11, while Figure 4.19 shows an example of checking the door gap and flushness on assembled cars [Perceptron n.d.].

Quality appraisal is a crucial process for ensuring the quality of final products delivered to customers. In addition, it is key to identifying quality problems during manufacturing and serving as a foundation for problem-solving and CI. However, quality appraisal does not inherently improve quality by itself. Number three of Deming's 14 principles for QM is to cease dependence on inspection for achieving quality, as discussed earlier in this chapter. From this perspective, the functionality of inspections serves for quality information.

According to Lean principles, quality inspections are normally considered non-value-adding activities from a customer's perspective. Therefore, it is important to assess the purpose, necessity, and cost of quality inspections for their justification. In certain situations, quality inspections are

Figure 4.19 Example of online quality inspection using laser sensors (Courtesy of Perceptron, Inc.).

mandated by government regulations or demanded by customers. In such cases, quality inspections can be categorized as necessary non-value-added; refer to subsection 8.4.1 of Chapter 8. In many cases, inspections are not needed when quality can be assured through built-in quality and error-proofing.

4.4.1.2 Layered Process Audits

Quality appraisals extend beyond inspecting the quality attributes of major WIP units and final products. In addition, quality inspections can focus on manufacturing processes to validate process settings and capabilities thereby preventing potential problems. One structured approach for process quality appraisals is called layered process audit (LPA) in operational management.

LPA confirms quality compliance at multiple levels within a manufacturing system, as illustrated in Figure 4.20, presenting a typical practice. In an LPA plan, team members conduct real-time audits, supervisors and managers perform periodic audits, and senior leadership conducts random audits.

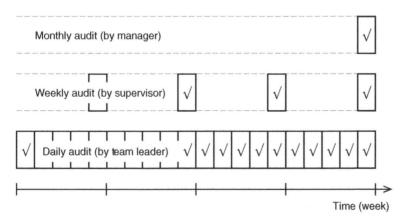

Figure 4.20 Illustration of layered process audit arrangement.

Each level of auditing has identified responsibilities and specific requirements for what and how to audit. Auditors use predesigned checklists and document their findings based on the audit criteria. Nonconformances can trigger corrective follow-up actions, particularly for issues without permanent solutions.

Initially introduced in the automotive industry during the 1990s, the LPA has a guideline [AIAG 2014] that can be customized to fit various manufacturing sectors. A study demonstrated the successful implementation of LPA in various manufacturing areas [Järnberg and Samuelsson 2022].

In addition to the discovery and validation provided by LPA, its effectiveness should be measured. For instance, LPA-related metrics include various aspects such as the percentage of audits completed on time, percentage of conformity, completion of corrective actions, and recurrence of nonconformances. Vigilantly monitoring these LPA-related metrics enables operations management to discern patterns, trends, and opportunities for improvement effectively.

4.4.1.3 Error-proofing

In many cases, both quality issues and non-quality problems are preventable and should be addressed during product design and manufacturing process planning. One effective prevention method is called error-proofing or mistake-proofing, which makes quality issues either impossible to occur or immediately detectable when they do occur.

An example of error-proofing is preventing a driver from leaving the headlights on. In the past, a reminder sign in a parking lot could be helpful (see Figure 4.21), but it was not preventive. In modern car designs, the headlights automatically shut off with a delay when the driver leaves the vehicle. This automatic function prevents forgetting-type mistakes.

Error-proofing practices can prevent quality issues, as demonstrated by the example in Figure 4.22 [Tang 2018]. In this case, a vision-sensor monitor with image processing can detect whether the bracket is in place, providing a warning function if the bracket is missing. Furthermore, both the bracket and its fixture are intentionally designed asymmetrical, ensuring the bracket can only be positioned in a specific way. These combined warning and prevention functions may guarantee the bracket is correctly positioned, eliminating any possibility of it being missed or placed incorrectly. A more effective approach is to use proximity sensors for the bracket, replacing vision checks, which has been widely implemented.

In many cases, it may not be practical to prevent all quality issues due to technical or financial constraints. In such situations, it is advisable to implement real-time monitoring to immediately

Figure 4.21 Example of error-proofing – warning sign.

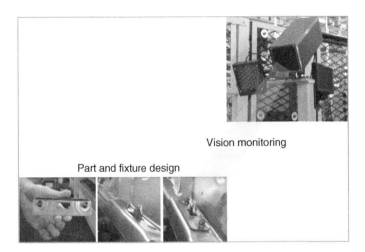

Figure 4.22 Example of error-proofing – prevention mechanism (Tang, 2017/with permission of SAE International).

detect issues, rather than waiting until the end of a manufacturing process. In addition, visual operational instructions can be helpful.

Subsection Summary: Quality appraisals, include inspections, process auditing, and tests, fulfill their functions in manufacturing. Proactive methods such as error-proofing, can be more effective to enhance quality and system performance.

4.4.2 Quality Improvement Tools

In manufacturing, various tools contribute to quality improvement, categorized based on their functions in root cause analysis. This subsection covers eight common tools and methods, each discussed with examples:

- Qualitative brainstorming and review (see Figure 4.23)
- Quantitative analysis for patterns (see Figure 4.24)
- Quantitative network methods for relationship (see Figure 4.25)

Numerous books and articles offer detailed discussions on these tools, along with additional options. Many other tools align with these three groups or even form new categories. Readers are encouraged to explore different and innovative ways to apply quality tools for CI.

4.4.2.1 Qualitative Tools for Brainstorming
The first category encompasses qualitative tools utilized for team brainstorming and reviews. These include the cause-and-effect diagram, relation diagram, and affinity diagram.

1. A cause-and-effect diagram: Also known as a fishbone diagram or Ishikawa diagram, this visual tool aids in the early stages of problem-solving and improvement projects during cross-functional team brainstorming. It facilitates understanding various factors and root causes contributing to a problem. Figure 4.23a illustrates six general contributing factors to an identified problem. Note that these factors may not be necessary for all projects, and there could be additional factors beyond the six listed. It is essential to focus on the primary factors when using this tool to establish a solid foundation for further analysis.

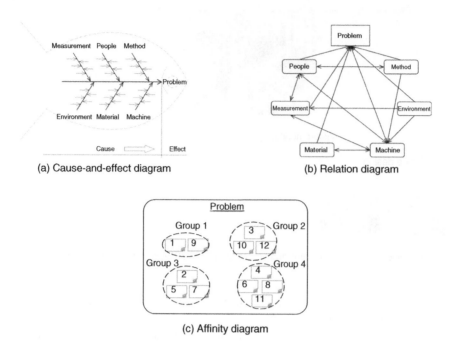

(a) Cause-and-effect diagram

(b) Relation diagram

(c) Affinity diagram

Figure 4.23 Qualitative tools for brainstorming and review.

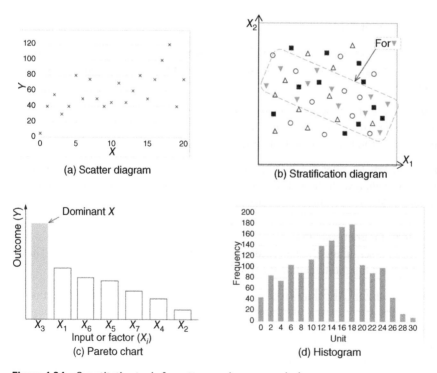

(a) Scatter diagram

(b) Stratification diagram

(c) Pareto chart

(d) Histogram

Figure 4.24 Quantitative tools for pattern and reason analysis.

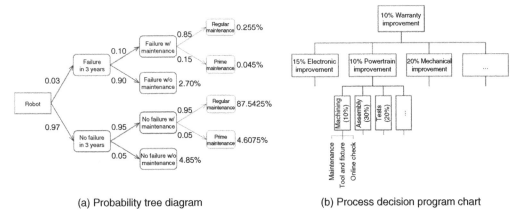

(a) Probability tree diagram   (b) Process decision program chart

Figure 4.25 Quantitative network methods for relationship analysis.

2. A relation diagram: (see Figure 4.23b) Also known as an interrelation or network diagram, this tool illustrates the interrelationship between the influencing factors. For complex problems, interrelationships between factors, such as those between people involved and methods used, jointly influence the problem. This is a useful feature compared to a cause–effect diagram. However, a relation diagram does not indicate multiple subfactors under a main factor. Therefore, using both a cause–effect diagram and a relation diagram simultaneously can be beneficial for a comprehensive problem analysis.

3. An affinity diagram: A tool used for organizing and consolidating substantial amounts of ideas and considerations into logical groups based on their natural relationships (Figure 4.23c). It is a mind-mapping technique used in brainstorming sessions to organize ideas or concerns into characteristic groups. This tool can help identify patterns and relationships among different pieces of information. For example, an affinity diagram can be used to organize ideas into groups, including initial hypotheses, potential avenues for root cause analysis, solution suggestions, and implementation planning.

These qualitative tools play a crucial role in the initial stages of problem analysis and solution development, providing a structured approach to understanding and addressing complex issues.

4.4.2.2 Quantitative Tools for Patterns

In most scenarios, quantitative analysis is essential not only for identifying factors and their relationships in root cause identification but also for assessing the impact of a problem and the benefits of the improvement project. This type of analysis is also crucial in prioritizing root causes for effective solutions. Four quantitative analysis tools are reviewed below:

1. Scatter diagram: A graphical tool that illustrates the relationship or pattern between two sets of numerical data in an X–Y plot (see Figure 4.24a). It visually represents correlations or patterns between two variables, such as a variable incrementing or decrementing with another variable, or between a factor and its effect with a quantified level of relationship. This tool can aid in the identification of cause-and-effect relationships and provide insights for further analysis.

2. A stratification chart (see Figure 4.24b): An analytical tool that sorts data into distinct categories, groups, or layers based on their similarities and patterns. This method is useful for revealing patterns of a specific variable in mixed data with multiple sources such as data from different

sources, equipment, departments, shifts, materials, and/or time periods. Stratification analysis provides graphical representations of grouped data, aiding in the identification of patterns and correlations hidden in mixed data.

3. A Pareto chart: A vertical bar graph that illustrates the relative significance of factors by ranking their influences normally from largest to smallest (refer to Figure 4.24c). In problem-solving, the dominant factor (e.g., X3 in the figure) should be given the highest priority for solutions. For example, Figure 3.10 in Chapter 3 includes four KPI charts in a Pareto format.

4. Histogram chart: (See Figure 4.24d) Displays the frequency distributions of data. The horizontal axis presents intervals or "bins," while the vertical axis shows the frequency of data value falling within each bin. Histogram charts can be used for various analyses, such as WIP distribution analysis to reveal the operational states of upstream and downstream systems. An application example is shown in Figure 3.25 in Chapter 3.

These quantitative tools play a crucial role in pattern analysis, reason identification, and decision-making, providing a structured approach for assessing and improving various aspects of a system.

4.4.2.3 Quantitative Network Tools

Quantitative network analysis is a cornerstone in problem-solving and CI, valued for its ability to comprehensively address multiple contributing factors and root causes involved in complex issues. Using these tools, operational management can gain valuable insights into process relationships and risks, enhancing data-driven decision-making.

1. Probability tree diagram: This tool combines network analysis and statistical analysis as a tool for quantitatively analyzing relationships (or risks) among primary factors or root causes (see Figure 4.25a). This diagram includes the probability of each element on a branch, aiding in the identification of fundamental issues and the prioritization of improvement projects.

2. Process decision program chart (PDPC) (See Figure 4.25b): Utilizes a tree diagram, PDPC serves as a planning tool to systematically identify potential issues and incorporate additional information. It is particularly useful for risk analysis to address potential issues before they occur. PDPC can also aid in prioritizing root causes using available data.

3. Quantification of qualitative methods: In addition to quantitative tools, some qualitative methods can be quantified by incorporating numerical ratings of relationships between multiple contributing factors and identified problems. For example, a cause-and-effect diagram may be presented in a matrix format, as shown in Table 4.5. The table provides an example with numerically rated relationships between multiple contributing factors (causes) and identified problems (effects).

These quantitative network methods play a pivotal role in relationship analysis, risk assessment, and decision-making, providing a systematic approach to addressing and improving complex issues.

Subsection Summary: A diverse array of tools aimed at problem-solving are reviewed, ranging from qualitative brainstorming methods to quantitative analysis. With their characteristics and purposes, these tools enable effective root cause identification for quality improvement.

Table 4.5 Example of cause-and-effect matrix.

10 = very strong correlation 7 = strong correlation 5 = some correlation 1 = no correlation Blank = pending	Output (effect) Subsystem A			
Input (cause)	Scrap rate	Repairability	Availability	Total
1 Method	6	5	3	14
2 Machine	8	5	7	20
3 People	6	8	3	17
4 Material	6	7	2	15
5 Measurement	7	4	5	16
6 Environment	8	8	2	18
Total	41	37	22	

Chapter Summary

4.1. Quality Management

1. Quality is subjective, defined by ASQ and ISO, with dimensions such as performance, safety, functionality, and reliability.
2. Quality drives operational excellence, impacting profit, productivity, and efficiency by reducing waste, minimizing disruptions, and ensuring stability.
3. Manufacturing quality can be measured by first-time quality, rework/repair, and the amount of scrap produced.
4. QMS has three pillars: quality planning, QC, and CI, with key elements of context, leadership, support, operation, performance evaluation, and improvement.
5. Several international and industrial standards exist for QMS implementation.

4.2. Cost Analysis of Quality

6. Quality cost has the following two major categories: the cost of good quality, which includes prevention and appraisal, and the cost of poor quality, which encompasses internal and external failures, or the PAF model.
7. Quality cost items, both direct and indirect, have network-like relationships in manufacturing.
8. Total quality cost combines the costs of good and poor quality. It can be at an economic (lowest) level, the best quality level, or somewhere in between.
9. Quality investment in improvement projects can be analyzed over different time periods with an interest rate for long-term cost-effectiveness.

10. Companies typically have an MARR (minimum acceptable rate of return) as a benchmark for judging project cost-effectiveness.

4.3. Quality in Production Throughput

11. Poor product quality negatively impacts operational performance, resulting in defective parts, repair work, and scrap, thereby requiring additional resources.
12. Quality directly affects OEE as its element. It also indirectly impacts OEE via availability and speed performance.
13. A manufacturing operation has its throughput yield (TPY) or first-time quality (FTQ) without repair.
14. The rolled throughput yield (RTY) assesses cumulative quality in multi-operation serial systems.
15. Quality in parallel systems has unique considerations due to the distributions of quality attributes from parallel subsystems.

4.4. Discussion of Improving Quality

16. Appraisal methods, including inspections, testing, analysis, and reporting, verify and ensure the quality of products, including WIPs and incoming materials, against specifications.
17. LPA (layered process audit) is an approach to validate process settings and compliance at multiple levels (team, supervision, and management) and frequencies.
18. Error-proofing, using preventive mechanisms or warning signals, is a proactive approach to ensure product quality.
19. Qualitative analysis tools for brainstorming include the cause-and-effect diagram, relation diagram, and affinity diagram.
20. Quantitative tools for patterns include the scatter diagram, stratification chart, Pareto chart, and histogram chart. For interrelationship, tools include the probability tree diagram and process decision program chart.

Exercise and Review

Exercises

4.1 The implementation of an automated in-process inspection necessitates a new investment of $40,000, aimed at reducing the cost of internal failures. Given an estimated weekly cost saving of $1500 from internal failures and assuming no other changes in costs, calculate the break-even point for this investment.

4.2 An improvement project requires an investment of $15,000 and is projected to yield a cost saving of $3500 per quarter due to improved product quality over the next two years. The company's minimum acceptable rate of return (MARR) is set at 14%. Determine whether this project is worth undertaking.

4.3 An improvement project requires an investment of $15,000, and the company's MARR is 14%. Calculate the minimum expected cost saving per quarter from improved product quality over two years, that would make this project worthwhile.

4.4 A manufacturing system consists of eight workstations in series, with TPY values of 0.99, 0.98, 0.95, 0.97, 0.98, 0.95, 0.98, and 0.99, respectively. Calculate the RTY of the entire system.

4.5 A system comprises three parallel, identical subsystems, each with mean values of a quality attribute of 15.2, 14.5, and 13.7, respectively. Estimate the mean value of this quality attribute for the combined products from all subsystems.

4.6 A system comprises two parallel, identical subsystems, each with similar mean values for a quality attribute but different variations. The standard deviations for this attribute are 5.5 and 7.3 for the two subsystems, respectively. Estimate the standard deviation value of this quality attribute for the combined products from both subsystems.

Review Questions

(The chapter covers these topics. For further discussion, it is recommended to seek additional information and examples. Diverse perspectives are encouraged.)

4.1 Define the quality of a product that you own or have worked on.

4.2 Furnish examples of product quality dimensions.

4.3 Construct a concise SIPOC matrix for a manufacturing operation.

4.4 Examine the relationships between the three pillars of a QMS.

4.5 Describe the definition of QMS based on a QMS standard and compare it with Juran's 3-pillar model.

4.6 Describe the four types of quality costs and their direct and indirect relationships to performance excellence in a manufacturing operation, providing examples.

4.7 How can throughput performance be integrated into total quality cost analysis? Provide an example.

4.8 It is often stated that economic quality is not sufficient. Do you concur? Illustrate with an example.

4.9 Compare a project evaluation based on a break-even analysis and an economic analysis.

4.10 For an economic evaluation of a five-year improvement project, would you suggest using a break-even analysis? Why?

4.11 Look for an example demonstrating how quality indirectly impacts operational throughput.

4.12 Explain how quality influences the availability (A) and performance (P) in OEE, with an example.

4.13 Clarify the concepts of throughput yield (TPY) and first-time quality (FTQ), and comment on the quality with the inclusion of repaired units (refer to Figure 4.11)

4.14 Examine the RTY of a serial system for its relevance in bottleneck identification.

4.15 Explain quality considerations in parallel manufacturing systems.

4.16 Explain the benefits and costs of quality inspections, providing an example.

4.17 Detail LPA (layered process audits) and provide an example.

4.18 Review a preventive error-proofing method with an example.

4.19 Review a type of quality tool for its suitability and limitations in quality improvement projects.

4.20 Identify another quality tool and assess whether it fits into a category mentioned in the text or propose a new category for the tool.

References

AIAG 2014. CQI-8 Layered Process Audit Guideline, Automotive Industry Action Group, ISBN: 9781605343006. Southfield, MI.

Alglawe, A., Schiffauerova, A. and Kuzgunkaya, O. 2019. Analysing the cost of quality within a supply chain using a system dynamics approach. Total Quality Management & Business Excellence, 30(15–16), pp. 1630–1653.

ASQ n.d. Quality Glossary. https://asq.org/quality-resources/quality-glossary. Accessed July 2020.

Butler-Bowdon, T. 2017. 50 Economics Classics: Your Shortcut to the Most Important Ideas on Capitalism, Finance, and the Global Economy, ISBN: 978-857886733, Nicholas: Brealey.

Deming, W.E. 1986. Out of the Crisis, 2nd ed., ISBN: 978-0262541152, MIT Press.

Farooq, M.A., Kirchain, R., Novoa, H. and Araujo, A. 2017. Cost of quality: evaluating cost-quality trade-offs for inspection strategies of manufacturing processes. International Journal of Production Economics, 188, pp. 156–166.

Fawzy, M.F. and Olson, E.W. 2018. Total quality management & APPLE success. In Proceedings of the American Society for Engineering Management 2018 International Annual Conference. 17–20 October 2018. Coeur d'Alene, Idaho, USA.

Forbes Insights 2017. The Rising Economic Power of Quality – How Quality Ensures Growth and Enhances Profitability. https://www.forbes.com/forbesinsights/asq_economics_of_quality/index.html. Accessed November 2022.

Franz, M. and Wartzack, S. 2022. Tolerance optimization of patch parameters for locally reinforced composite structures. Applied Composite Materials, pp. 1–24.

Herzog, I. and Grabowska, M. 2021. Quality cost account as a framework of continuous improvement at operational and strategic level. Management and Production Engineering Review, 12(4), pp. 122–132.

IATF 2016. IATF 16949 Quality Management System Requirements for Automotive Production and Relevant Service Parts Organizations, International Automotive Task Force.

ISO 2015a. ISO 9000:2015 Quality Management Systems — Fundamentals and vocabulary, Geneva, Switzerland.

ISO 2015b. ISO 9001:2015 Quality Management Systems — Requirements, International Organization for Standardization, Geneva, Switzerland.

Janatyan, N., Shahin, A. and Khodaparastan, M. 2023. Optimising the costs of quality combination in the product life cycle by Taguchi design of experiments. International Journal of Productivity and Quality Management, 39(1), pp. 20–41.

Järnberg, F. and Samuelsson, M. 2022. Integration of lean management system into the digitalized factory - an investigation into the use of an IT-tool for layered process audits and leader standard work (Master's thesis in Quality and Operations Management, Chalmers University of Technology).

Juran, J.M. 1986. The quality trilogy—a universal approach to managing for quality, In ASQC 40th Annual Quality Congress in Anaheim, California, 20 May 1986.

Li, Y., Zhang, F.P., Yan, Y., Zhou, J.H. and Li, Y.F. 2019. Multi-source uncertainty considered assembly process quality control based on surrogate model and information entropy. Structural and Multidisciplinary Optimization, 59, pp. 1685–1701.

Liang, B., Zhou, T. and Hu, Y. 2022. Research and innovative methodology on closed-loop analysis & improvement mechanism based on cost of quality system. International Journal of Information and Management Sciences, 33, pp. 167–182.

Martin, J., Elg, M. and Gremyr, I. 2020. The many meanings of quality: towards a definition in support of sustainable operations. Total Quality Management & Business Excellence, pp. 1–14.

Moradinaftchali, V., Song, L. and Wang, X. 2016. Improvement in quality and productivity of an assembled product: a riskless approach. Computers and Industrial Engineering, 94, pp. 74–82.

Obied-Allah, F. 2016. Quality cost and its relationship to revenue sharing in supply chain. Accounting and Finance Research, 5(3), pp. 173–189.

Owen, J.H. and Blumenfeld, D.E. 2008. Effects of operating speed on production quality and throughput. International Journal of Production Research, 46(24), pp. 7039–7056.

Perceptron n.d. Optimize fit and finish of adjacent panels or trim assemblies. https://perceptron.com/solutions/gap-and-flush/. Accessed January 2022.

Plewa, M., Kaiser, G. and Hartmann, E. 2016. Is quality still free? Empirical evidence on quality cost in modern manufacturing. The International Journal of Quality and Reliability Management, 33(9), pp. 1270–1285.

Putri, N.T., Yusof, S.M., Hasan, A. and Darma, H.S. 2017. A structural equation model for evaluating the relationship between total quality management and employees' productivity. The International Journal of Quality and Reliability Management, 34(8), pp. 1138–1151.

Rahardjo, W.H., Farizal, F. and Gabriel, D.S. 2020. Cost of quality system in passenger car plant: a methodology of implementation. In IOP Conference Series: Materials Science and Engineering, Vol. 909, No. 1, p. 012069. IOP Publishing.

Ruediger, P., Claus, F., Hamann, B., Hagen, H. and Leitte, H. 2021. Combining visual analytics and machine learning for reverse engineering in assembly quality control. Electronic Imaging, 2021(1), pp. 60405-1–60405-13.

Samawi, G.A., Abu-Tayeh, B.K., Yosef, F., Madanat, M. and Al-Qatawneh, M.I. 2018. Relation between total quality management practices and business excellence: evidence from private service firms in Jordan. International Review of Management and Marketing, 8(1), pp. 28–35.

Schwerha, D., Casey, A. and Loree, N. 2020. Development of a system to integrate safety, productivity, and quality metrics for improved communication and solutions. Safety Science, 129, p. 104765.

Shingo, S. 1986. Zero Quality Control: Source Inspection and the Poka-Yoke System, ISBN: 978-0915299072, CRC Press.

Tang, H. 2017. Automotive Vehicle Assembly Processes and Operations Management, ISBN: 978-0-7680-8338-5, Warrendale, PA: SAE International.

Tang, H. 2018. Manufacturing System and Process Development for Vehicle Assembly, ISBN: 978-0-7680-8346-0, Warrendale, PA: SAE International.

Tang, H. 2022. Quality Planning and Assurance – Principles, Approaches, and Methods for Product and Service Development, ISBN-13: 978-1119819271, Hoboken, NJ: Wiley.

5

Maintenance Management and Throughput

Equipment maintenance plays a dual role: maintaining equipment in good working order and supporting manufacturing operations and throughput. From this perspective, this chapter examines maintenance principles and practices based on the purposes, referred to as throughput-focused maintenance.

5.1 Maintenance Principles

5.1.1 Reliability Principles

5.1.1.1 Equipment Reliability

Reliability is defined as the probability that a piece of equipment will consistently perform its intended functions without failure for a specific length of time. In manufacturing, it signifies the likelihood that a system can operate without failure over a specified period. As such, reliability plays a crucial role in achieving desired manufacturing throughput.

A key concept related to reliability is the equipment's failure rate (λ). The failure rate is defined as the number of failures occurring during a given interval, for instance, per hour or per cycle.

$$\lambda = \frac{\text{Number of failures}}{\text{Total time}}$$

To estimate a machine's reliability, the calculation uses a simple exponential formula based on the failure rate. Reliability $R(t)$ as a function of time is calculated as follows:

$$R(t) = e^{-\lambda t}$$

If the failure rate is known and is a constant, calculating estimating equipment reliability becomes straightforward. For instance, if a robot has a failure rate of 0.0001 failures per hour, its reliability for eight hours is:

$$R(8) = e^{-0.0001 \times 8} \approx 99.9\%$$

The result means that the robot's reliability is approximately 99.9% over an eight-hour period.

This estimation is conceptual as it relies on the assumptions of a constant failure rate and a single failure mode. However, modern machines are complex and can exhibit multiple failure modes with varying failure rates.

Manufacturing System Throughput Excellence: Analysis, Improvement, and Design, First Edition. Herman Tang.
© 2024 John Wiley & Sons, Inc. Published 2024 by John Wiley & Sons, Inc.
Companion website: www.wiley.com/go/Tang/ManufacturingSystem

5.1.1.2 Failure Effect Severity

In addition to the failure rate, the severity of a failure serves as a useful reference for prioritizing maintenance work. The severity of a failure can be assessed based on various factors, including frequency, average duration, and consequences within a specific time, e.g., a week. Consequences can be rated based on their impact, such as operations, safety, cost, and environment, using a scale from 1 to 10. It is essential to establish the consequence rating before conducting a severity analysis.

To calculate the severity (S) of individual failures, denoted as S_j, using the following formula:

$$S_j = F_j \times D_j \times C_j$$

Here, F_j represents the frequency, D_j represents the average duration, and C_j represents the consequence rating of failure "j." Note that the calculated severity (S_j) is comparative within the same or similar operations. The elements F_j, D_j, and C_j should use the same units for all failures.

For instance, a system experiences several failures during its one-week production period, as shown in Table 5.1. In the table, Failures 1 and 3 have high-severity values and should be addressed first.

As the severity score is a composite metric from three elements, it is worth checking the individual elements for a high-severity score for a better understanding of contributing factors. In addition, using a single severity score can be a quick reference for prioritization but may be inaccurate because it does not encompass other factors such as possible dependencies between failures.

5.1.1.3 Failure Rate Curves

Equipment reliability follows a certain pattern over time, which is often depicted using a bathtub-shaped curve. A general representation of this curve is shown in Figure 5.1, and it can exhibit different shapes [Jiang 2013, Tóth and Jónás 2014]. Good maintenance practices can extend the useful life of most mechanical equipment.

This bathtub-shaped curve is commonly observed in most mechanical components and equipment. However, electrical and electronic components and products do not typically follow the bathtub-shaped curve according to the literature [Gaonkar et al. 2021].

The discussion of reliability and its pattern extends to a manufacturing system. It comprises many mechanical and electrical components including tooling, fixtures, equipment, conveyors, controllers, power units, and sensors. As a result, the reliability of a manufacturing system is significantly more complex than that of individual equipment.

In addition to its complexity, a manufacturing system's reliability is affected by a wide range of factors, including individual components' reliability, the interactions between components, maintenance, and the system design. Therefore, when operating and maintaining a manufacturing system, it is crucial and challenging to assess the reliability of all components and their interactions for operational performance.

Table 5.1 Example of downtime severity for maintenance planning.

Failure j	Frequency F_j	Duration D_j	Consequence C_j	Severity S_j
1	7 times	10 minutes	7 (downtime)	490
2	15	5	5 (quality)	375
3	2	25	9 (safety)	450
4	32	0.5	4 (slowness)	64
...				

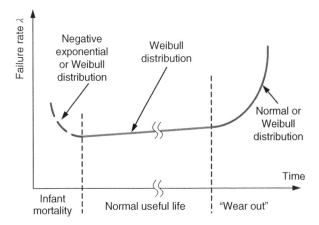

Figure 5.1 Conceptual reliability "bathtub" curve.

5.1.1.4 Failure Data Characteristics

Equipment failures exhibit distinct characteristics, and common probability distributions (density functions) used to describe failure data include the Normal distribution, negative exponential distribution, and Weibull distribution.

Figure 5.2 illustrates these three types of probability distributions as having varying shapes and locations based on their parameters. To simplify an analysis, it is typically assumed that equipment reliability follows one of these distributions or a combination, as seen in the curve in Figure 5.1. If the failure rate of equipment or system is known, the corresponding probability distribution function and curve can be derived and developed.

Due to various factors in manufacturing, maintenance, and/or design of the equipment, the behavior of failures, such as their timing and frequency, can appear irregular and complicated. That is because these failures may result from a combination of different data distributions. In maintenance practices, addressing the root causes of failures, their timing, repairs, and prevention is often more important than dealing with distribution characteristics.

Subsection Summary: Understanding reliability principles, analyzing failure severity, and studying distribution patterns provide insights into optimizing maintenance practices in complex manufacturing systems, leading to throughput excellence.

5.1.2 MTBF and MTTR

5.1.2.1 Concept of MTBF and MTTR

Mean time between failures (MTBF) and mean time to repair (MTTR) are two key reliability-related KPIs within the context of system throughput and maintenance performance. From their prior introduction, this subsection reviews them in more detail in the context of reliability and maintenance. Figure 5.3 illustrates a scenario involving MTBF and MTTR.

MTBF represents the average time between the failures of a piece of equipment or a system and is typically measured in hours. It is calculated using the following formula:

$$\text{MTBF} = \frac{\text{Total available time}}{\text{Number of failures}}$$

On the contrary, MTTR represents the average time required to repair a failed piece of equipment or a system and return it to full functionality. It is also referred to as mean time to recovery or mean

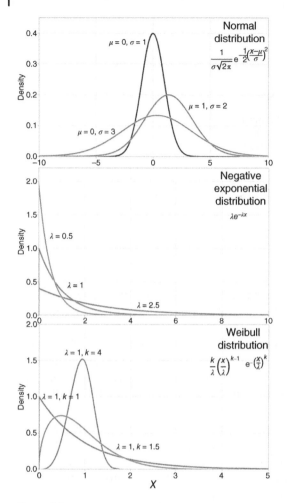

Figure 5.2 Curves of Normal, Exponential, and Weibull distribution functions.

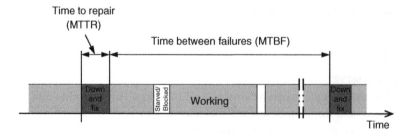

Figure 5.3 MTBF and MTTR of system and equipment.

time to restoration (MTTR). MTTR serves as a critical indicator of repair work efficiency, reflecting how quickly issues are resolved. The calculation for MTTR is as follows, often measured in minutes in mass production:

$$\text{MTTR} = \frac{\text{Total down time}}{\text{Number of downtime or repair}}$$

It is important to note that both MTBF and MTTR present average values over an extended period. The "M" in each acronym stands for arithmetic mean. To provide reliable insights, they require data collected over a sufficiently long period, such as one month.

MTBF and MTTR are crucial metrics for evaluating the efficiency of maintenance processes, thereby significantly impacting system operational availability.

5.1.2.2   Calculation of MTBF and MTTR

For instance, a manufacturing shop operates for 347 hours over a month with a two-shift production schedule. During this period, the shop experienced 54 failures with a total downtime of 190 minutes. The MTBF and MTTR can be calculated as:

$$\text{MTBF} = \frac{\text{Total available time}}{\text{Number of failures}} = \frac{347 - \dfrac{190}{60}}{54} \approx 6.37 \text{ (hours or 6 hours 22 minutes)}$$

$$\text{MTTR} = \frac{\text{Total down time}}{\text{Number of downtime or repair}} = \frac{190}{54} = 3.5 \text{ (minutes)}$$

By comparing these values to historical data and target values, one can assess the performance and reliability of the manufacturing shop's equipment and maintenance efficiency.

It is important to note that the MTBF of equipment is predetermined by its inherent reliability, influenced by various design factors. These factors include fault detectability, setups, repair accessibility, layout design, and equipment modularity. Once machines are in production, significantly improving their inherent maintainability can be challenging.

MTTR is a vital indicator of maintainability, reflecting the efficiency of troubleshooting, repair work, and functionality restoration. The influencing factors include task prioritization, spare availability, and the proficiency of maintenance technicians. Maintenance may improve MTBF by well managing these factors on the production floor. More exploration of these topics follows in later subsections.

5.1.2.3   Relationship between λ and MTBF

In addition to being expressed in failures per unit of time, the failure rate (λ) is inversely proportional to MTBF. Thus, λ can be estimated with a known MTBF using the formula:

$$\lambda = \frac{\text{Number of failures}}{\text{Total time}} = \frac{1}{\text{MTBF}}$$

Note that λ is calculated based on the total time, while MTBF calculation is typically based on the total available time (uptime). The relationship is accurate if the total time and total available time are about the same. If the two time periods are different, this relationship is a rough estimate or overall trend.

For example, if the MTBF of a manufacturing system is 16.7 hours, then the failure rate can be estimated as:

$$\lambda = \frac{1}{\text{MTBF}} = \frac{1}{16.7} \approx 0.06 \text{ (failures per hour)}$$

Using the same equation, MTBF can be estimated based on a known λ. For instance, if a piece of equipment has a constant $\lambda = 0.0005$ failures per hour over an extended period, its MTBF can be calculated as follows:

$$\text{MTBF} = \frac{1}{0.0005} = 2000 \text{ (hours)}$$

Note that this calculated MTBF is a mean value as an overall estimate.

Based on the relationship between λ and MTBF, the reliability for a defined period (t) can be expressed:

$$R(t) = e^{-\lambda t} = e^{-\frac{t}{\text{MTBF}}}$$

For instance, if a particular type of robot has an MTBF of 20,000 hours, one can estimate its reliability for running eight hours as follows:

$$R(8) = e^{-\frac{8}{20,000}} \approx 99.96\%.$$

This example provides reference information about the probability that the robot will operate without failure during the specified period.

5.1.2.4 Discussion on Failure Rate

An important characteristic of the failure rate is its additivity. If a system consists of (n) independent subsystems, the failure rate of the entire system (λ_{system}) is calculated as the sum of the individual failure rates (λ_i) of all subsystems:

$$\lambda_{\text{system}} = \sum_{i=1}^{n} \lambda_i$$

This additive property simplifies the assessment of the overall failure rate for complex systems composed of independent components. However, the independence assumption may not always be valid. Consequently, when calculating the system's failure rate based on the failure rates of its subsystems, the result is an approximation rather than an exact value.

Understanding the relationship between failure rate and operation time reveals how reliability tends to change with operation time. Reliability typically declines as operation time increases, often experiencing a rapid drop after a certain point, such as six months, as shown in Figure 5.4.

This information is important for maintenance management for its maintenance activity planning. In addition, the overall trend provides an insight into when doing major repairs or replacements is necessary as reliability is no longer acceptable after a certain period.

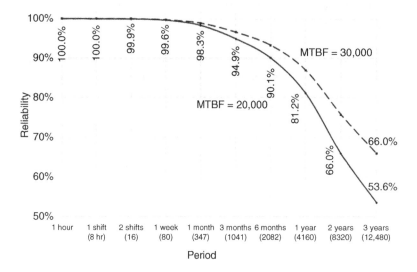

Figure 5.4 Reliability trends under different MTBF values over time.

Subsection Summary: As key metrics for maintenance and system performance, MTBF, MTTR, failure rate, and their relationships are studied in this subsection. Understanding these metrics and associated principles provides valuable guidance for effective maintenance planning and improvements.

5.1.3 Role of Maintenance in System Throughput

5.1.3.1 Maintenance in Production Management

The discussion of maintenance starts with its role and objective in manufacturing. Production is at the heart of a manufacturing plant, and maintenance serves as essential support for production operations. Figure 5.5 depicts this relationship, highlighting the crucial role of maintenance in achieving excellent production performance from a plant management perspective.

In many manufacturing plants or organizations, production management and maintenance management are organized as separate departments. In terms of their functional relationship, they can be likened to customer–supplier partners, as illustrated in Figure 5.6.

Both production and maintenance departments play pivotal roles in reducing equipment downtime duration and frequency. Therefore, they must work together to achieve high operational availability and continuous throughput improvement.

Choosing the optimal maintenance strategy and practice is a complex task, requiring good balancing of various, interconnected factors. These factors include equipment reliability, maintenance cost, resource availability, and throughput performance. For instance, scheduling maintenance during nonproduction hours can minimize operational disruptions, a critical consideration when evaluating maintenance strategies. Maintenance strategies and their characteristics are discussed in the next section 5.2.

5.1.3.2 General Relation to Availability

Building on the understanding of maintenance's importance to operational performance, the impact of equipment maintenance on operational availability can be further explored. The impact is affected by various factors including equipment type, age, design issues, and usage patterns. For

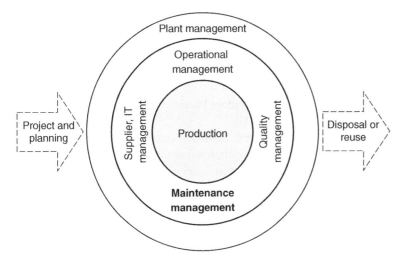

Figure 5.5 Role of maintenance management in a plant management system.

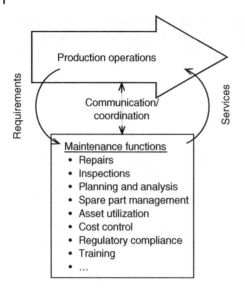

Figure 5.6 Customer-supplier like the relationship between production and maintenance.

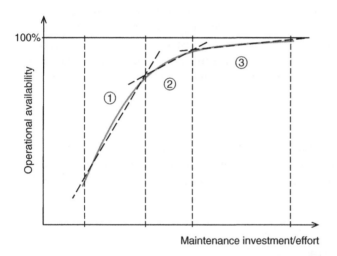

Figure 5.7 General relationship between availability and maintenance.

instance, certain equipment design issues may lead to more breakdowns during the initial years before becoming stable over time [Hagström et al. 2020].

For discussion purposes, Figure 5.7 depicts a general, nonlinear relationship between maintenance investment (such as time, cost, workforce) and operational availability. This overall trend is true over an extended period and can serve as a reference in maintenance management.

This nonlinear relationship can be segmented into three zones to analyze the relationship, each characterized by estimated linear slopes. In Zone ①, operational availability is low, increased maintenance efforts can notably increase availability. Conversely, Zone ③ indicates a scenario where availability is already high, and additional maintenance efforts show diminishing returns. Zone ② strikes a balance, representing a balanced proportion between maintenance efforts and operational availability, which may be the cost-effective, desired situation in many cases.

With an understanding of this relationship, a fundamental question arises: How can we know and choose the "optimal" maintenance level to achieve the desired system performance while considering total maintenance cost and resource allocation? This topic will be explored further in this chapter.

In addition, beyond operational availability, maintenance activities play a pivotal role in ensuring product quality, employee safety, equipment lifespan, efficient utility usage, and more. The principles and processes on throughput can be applied to these areas with certain adjustments.

5.1.3.3 Quantitative Maintenance Impact

The relationship between maintenance efforts and operational availability can be quantitatively assessed. Without considering other factors and variations, the operational availability (A) of a manufacturing system can be estimated using the following equation:

$$A = \frac{\text{Uptime}}{\text{Uptime} + \text{Downtime}} = \frac{\text{MTBF} \times n}{(\text{MTBF} + \text{MTTR}) \times n} = \frac{\text{MTBF}}{\text{MTBF} + \text{MTTR}} = \frac{\text{MTBF}}{t}$$

In this equation, "n" represents the number of downtime events during a given period "t." This equation builds a direct relationship between A and both MTTR and MTBF.

From the equation, it can be derived that availability (A) is proportional to MTBF and proportional to 1−MTTR. The relationships are summarized in Table 5.2, where the symbol $\propto$ indicates proportionality.

Figure 5.8 provides an example of these relationships, illustrating them in the context of an operation with 7.5 hours of working time ($t = 7.5$) and five downtime events ($n = 5$). It is important to note that the figure shows an overall trend as the slopes and scales of the relationships vary with MTBF and MTTR.

Table 5.2 Relations between A and MTTR and between A and MTBF.

	MTBF	MTTR
A	$A = \dfrac{\text{MTBF}}{t}$ i.e., $A \propto \text{MTBF}$	$A = 1 - \dfrac{\text{MTTR}}{t}$ i.e., $A \propto 1 - \text{MTTR}$

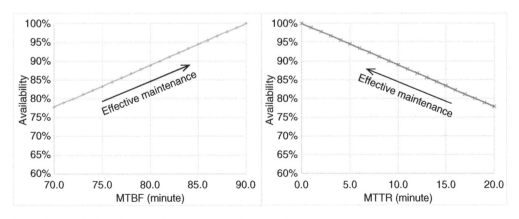

Figure 5.8 Relations between A and MTTR and between A and MTBF.

The figure clearly indicates that effective maintenance directly contributes to operational availability, by increasing uptime (MTBF) and/or reducing downtime (MTTR). The overall trend remains true, while the slopes and scales vary with MTBF and MTTR. The actual relationship may form a complex curve when considering other factors and variations, as shown in Figure 5.7.

5.1.3.4 Quality Maintenance for Throughput

In addition to the relationship between maintenance to availability, maintenance work also plays a pivotal supporting role in quality assurance and continuous improvement. Figure 5.9 provides an overall picture of how maintenance work directly impacts throughput performance and indirectly influences via quality.

Chapter 4 discusses that only good products are counted toward production throughput, while defective products need to be repaired or scrapped. On this, good maintenance keeps manufacturing processes in an optimal capability and thus minimizes defective parts produced. In addition, as discussed in Chapter 2, quality is one of the three elements in OEE. Good quality directly contributes to OEE.

In the light of maintenance's dual impacts on availability and quality, planning maintenance for both becomes essential. Some studies focused on joint control of maintenance and quality [Bouslah et al. 2018]. A starting point is to study the quantitative relationships between maintenance and availability and between maintenance and quality, as conceptually illustrated in the relationship curves in Figure 5.9. Subsequently, the priority of maintenance tasks can be determined.

A trial experiment can be used to quantify the relationship. The first step involves KPIs for both sides such as the maintenance workforce and quality defect rate. These variables can be converted into monetary values, as discussed previously. In the second step, plan and execute three levels of maintenance over a defined period such as two weeks. The third step involves observing maintenance efforts and product defect rates to estimate their relationship.

This relationship-building process, when completed, establishes a quantitative relationship. It is important to note that this task requires extra effort, dedicated teamwork, and management approval. Quantifying these relationships can guide data-driven decisions, in addition to relying on experiential knowledge.

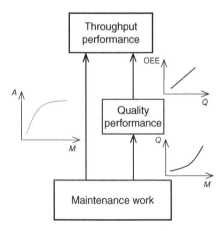

Figure 5.9 Maintenance contributions to throughput directly and via quality improvement.

Subsection Summary: Maintenance is a key supporting function in manufacturing management for system operations. Recognizing the relationships, particularly quantitatively, between maintenance and availability, as well as maintenance and quality, can make maintenance effective for throughput excellence.

5.2 Equipment Maintenance Strategies

5.2.1 Overall Maintenance Management

5.2.1.1 Maintenance Planning

Transitioning from general maintenance principles, maintenance plays a pivotal supporting role and begins with planning. According to ISO 55001 Asset Management [ISO 2014], Maintenance management should be an integral part of the asset lifecycle management of business operations, as illustrated in Figure 5.10.

Asset lifecycle management provides guidance to achieve an optimal balance among system performance, financial soundness, equipment utilization, and resource utilization. Meantime, equipment lifecycle investment management focuses on the equipment used in production. Maintenance departments should follow the guidance and information provided by these two business functions to develop plans and execute planned actions. This underscores that maintenance planning and activities should align with overall business objectives while considering available resources.

As a crucial supporting function, a maintenance system coordinates with operational management on planning and execution (Figure 5.11). Many factors, such as operation schedules, the significance of machinery, the complexity of maintenance tasks, the availability of technology, and resources, are considered to support manufacturing operations.

In this dynamic relationship, both maintenance management and production management should collaborate as partners in the identification and prioritization of maintenance tasks.

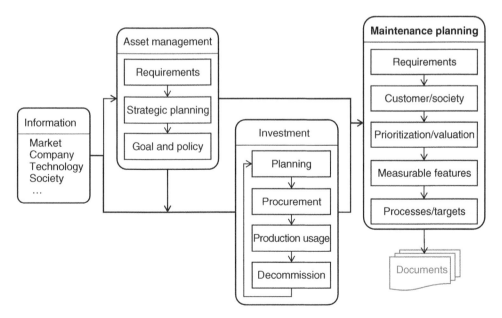

Figure 5.10 Process flow and factors of maintenance planning.

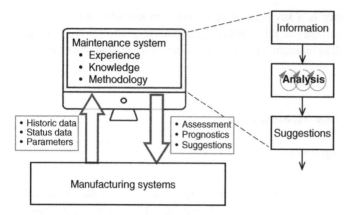

Figure 5.11 Integrated maintenance system.

This collaborative work should adhere to a data-driven process to make informed decisions in maintenance services. Several models and frameworks, like those proposed by Sala et al. [2021], address these key points.

As an integral part of maintenance, the quantity and storage location of spare parts are critical for reducing the system's MTTR. They should be determined based on several factors including the expected frequency of failures, requirements of maintenance, supply lead time, and cost. To strike a balance among these factors, spare parts can be classified into the following three types:

- Needed for recurrent failures: These spare parts are essential and should be readily available, located in proximity to the equipment sites.
- Necessary but for nonrecurrent situations: These parts are also essential, but the inventory level may be lower, contingent on utilization frequencies and supply lead time.
- Standard: These parts are typically in small quantities, can be stored near the related equipment, with more stock available in the storage area serving the entire system.

For high-volume production, even a short downtime can be costly, so minimizing MTTR is more important than spare part costs. The quantity of spare parts is also determined by delivery time if not stored on-site. Studies, such as [Antosz and Ratnayake 2019], focus on spare part management to support the increased availability of production systems.

5.2.1.2 Basic Maintenance Strategies
Maintenance strategies can be categorized into several types based on methods, arrangement, characteristics, etc. Figure 5.12 illustrates an overall categorization of three fundamental maintenance strategies.

As shown in Figure 5.12, the three types of maintenance strategies are (refer to Table 5.4 in the next section 5.3 on their characteristics):

- Corrective maintenance: It involves reactive responses to restore functions after failures have occurred or performance has failed to meet specifications. It aims to bring equipment functionality back to its optimal state. A typical example is repairing a faulty machine. Most corrective maintenance occurs during operations, leading to significant production losses.
- Preventive maintenance: It is conducted according to a predetermined schedule based on historical data and reliability analysis. Common examples include routine lubricating, inspection,

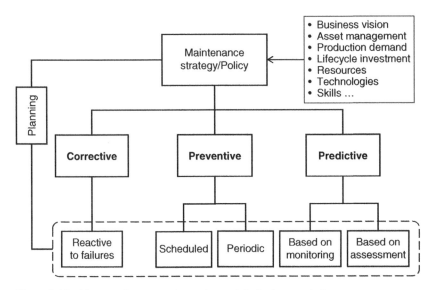

Figure 5.12 Three maintenance strategies and their characteristics.

calibration, and part replacement. While this strategy can reduce the likelihood of failures, it cannot eliminate them. It can minimize the impact of failures on operations.

- Predictive maintenance: This strategy utilizes techniques to continuously assess the conditions of equipment, identify early signs of potential problems, and evaluate their risks and severity. Predictive maintenance activities are triggered when imminent problems are detected. A simple example is that vibration sensors on electric motors can tell a motor's working condition and provide alerts, indicating potential problems. Thanks to its predictive nature, the maintenance work can be highly effective without unnecessary effort.

It is a common practice to use different maintenance strategies in a single maintenance system simultaneously. The proportions of three types of maintenance practices serve as indicators of maintenance effectiveness. For example, the breakdown of maintenance types is suggested [DOE 2010]:

- <10% reactive maintenance
- 25% to 35% preventive maintenance
- 45% to 55% predictive maintenance

A high percentage of preventive maintenance, particularly predictive maintenance, indicates a strong commitment to reliability principles, implying a high level of effectiveness in supporting operations. Implementing more preventive and predictive maintenances can be challenging. A study conducted in the Swedish aerospace industry showed that certain factors hindered the shift from a ratio of approximately 70% reactive and 30% preventive efforts to the opposite proportion [Sandberg et al. 2014].

Maintenance strategies and practices evolve with new industry developments such as Industry 4.0, vertical integration for information, asset-focused or total-cost-oriented approaches. A prominent trend is the shift toward more data-driven, predictive maintenance.

It is important to emphasize that equipment maintenance should not rest solely on the maintenance department. All organizational departments related to equipment should collaborate in addressing maintenance issues to optimize equipment utilization and system performance. This

Table 5.3 Tasks of manufacturing personnel in autonomous maintenance.

Maintenance technicians	Production workers
To analyze performance and breakdowns	To monitor equipment conditions
To conduct major repairs	To maintain "basic" machinery conditions
To plan and conduct preventive maintenance	To inspect and detect problems
To implement improvements	To propose/implement simple improvements

collaborative approach collaborative for total productive maintenance (TPM) is to be discussed in the next subsection 5.2.1.3.

5.2.1.3 TPM Approach

TPM represents a systematic strategy for maximizing manufacturing productivity, as evident in the "P" of its name. This "Total" aspect emphasizes its inclusion of all workers and considering various aspects of operations including throughput, quality, cost, and equipment life.

TPM is comprehensive, combining technical tasks, managerial actions, corporate procedures, cross-functional teams, and system life cycle considerations for optimal overall (operational and financial) effectiveness.

One essential aspect of TPM is autonomous maintenance (AM). AM assigns frontline workers the responsibilities of performing routine upkeep tasks, such as lubricating, inspecting, adjusting, and minor repairs, which go beyond traditional production duties. As shown in Table 5.3, a range of routine maintenance tasks are conducted by production workers along with maintenance technicians.

To perform the AM tasks assigned to production workers, specific authorizations, guidelines, and training are essential. Note that the responsibilities of production workers and maintenance technicians may overlap, requiring collaboration between the two teams.

AM implementation and success are closely related to business procedures and culture, fostering a sense of ownership over equipment care among production personnel and motivating them to enhance system performance.

Successful AM implementation can lead to significant performance improvement. For instance, a case study conducted at a textile plant shows that AM reduced minor stoppage by 29% and breakdown by 62%, resulting in an OEE increase from 75.09% to 86.02% [Ahmad et al. 2018]. Beyond productivity, TPM practices can enhance quality, cost efficiency, safety, and employee morale.

Subsection Summary: Maintenance strategies differ in their effectiveness. While corrective maintenance addresses breakdowns, preventive and predictive approaches aim to mitigate them through scheduled upkeep and condition monitoring. TPM takes it further, emphasizing cross-functional collaboration and focusing on enhancing system performance.

5.2.2 Reliability-Centered Maintenance (RCM)

5.2.2.1 Reliability Aging Pattern

Preventive maintenance, a common practice, involves overhauling or replacing components based on predetermined schedules. However, studies have found that this approach was not always cost-effective, as only approximately 11% of the components exhibited failure characteristics related to usage time [Nowlan and Heap 1978; Siddiqui and Ben-Daya 2009]. Since component

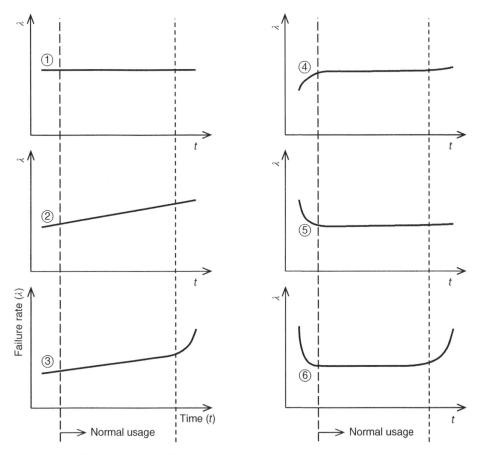

Figure 5.13 Six common reliability-time patterns.

failure characteristics are case-specific, the researchers identified six common reliability aging patterns, as illustrated in Figure 5.13.

Key points about the reliability-time characteristics include the following:

- Design influence: The characteristic curve of a component depends on its design, including the materials used and the part type.
- Trend variation: Curves ①, ④, and ⑤ remain relatively constant over time, while the others exhibit some degree of degradation.
- Component types: The general bathtub curve (curve ⑥) is often applicable to mechanical components, while many electronic components exhibit a characteristic like curve ③. Software failure rates may exhibit a gradual decrease.
- Beyond these, other patterns exist [Roesch 2012]: The actual reliability curve of a piece of equipment is a combination of these basic patterns, as it consists of various components.
- These reliability patterns assume normal operating conditions and proper maintenance.

Understanding the reliability of individual components is essential for comprehending potential failure modes of equipment, processes, and work cells. The understanding enables maintenance prioritization, focusing on high-impact items and addressing system bottlenecks. These topics will be further discussed in Chapter 7.

Therefore, analyzing reliability-time patterns plays a pivotal role in cost-effective maintenance planning. It helps optimally balance between replacing components too frequently and waiting too long. The incorporation of reliability in maintenance management has led to the development of reliability-lefted maintenance, which will be the subject of the next subsection 5.2.2.2.

5.2.2.2 Process of RCM

Reliability-lefted maintenance (RCM) is a maintenance principle and strategy that originated in the air transport industry. It focuses on ensuring the reliability of equipment, structures, and facility assets. RCM is designed to ensure that individual assets are matched with appropriate maintenance techniques to maximize cost-effective equipment uptime and enhance overall reliability. Figure 5.14a shows the RCM process, finding unreliable functions and root causes to support reliable operations. Throughput-focused maintenance shares similar workflows with RCM, as shown in Figure 5.14b. The main difference is in their focus, the first step in the workflow charts.

In the RCM process, the key is to evaluate functions to identify those with potential reliability issues and corresponding failure modes. These failure modes are then categorized into four types based on their consequences, determining the priority for scheduling and performing maintenance tasks.

1. Safety and environmental consequences
2. Operational consequences
3. Hidden failure consequences
4. Nonoperational consequences

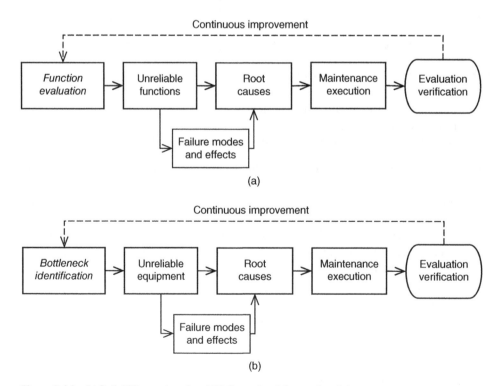

(a)

(b)

Figure 5.14 (a) Reliability-centered and (b) throughput-focused maintenance.

RCM practices should follow industry standards such as the SAE standard JA-1011 outlining evaluation criteria for RCM processes. The standard suggests asking seven questions when making maintenance decisions [SAE 2009]:

1. What is the desired performance level of the asset or equipment?
2. In what ways can this equipment fail?
3. What causes each failure?
4. What happens when each failure occurs?
5. Why does each failure matter?
6. What tasks should and can be performed to prevent failures?
7. What should be done if there is no suitable preventive task?

Similarly, the IEC has established a series of standards for dependability management. Part 3 of IEC 60300-3-11 is an application guide for RCM [IEC 2009]. Developed for aviation, these standards offer guidance for applying RCM analysis techniques to manage equipment and structure failures.

5.2.2.3   Considerations for RCM Implementation

Although RCM aims for reliable operations, its top priority may not always be productivity. In contrast, throughput-focused maintenance targets productivity improvements by addressing bottlenecks often caused by unreliable machines. Overall, by ensuring equipment reliability, the RCM approach can indirectly contribute to better throughput management and performance.

RCM can be applied collectively with the widely practiced preventive maintenance strategy and is more aligned with predictive maintenance practices. RCM can guide both strategies by providing more data-driven, reliability-focused information, enhancing their effectiveness. For instance, the information from RCM analysis on unreliable functions and corresponding root causes helps maintenance management plan and prioritize tasks.

RCM identifies and manages the risks of functional failures to prevent their occurrence. Given its risk-focused approach, RCM shares similar principles and processes with failure mode and effect analysis (FMEA), a systematic, proactive approach [Tang 2022]. While RCM and FMEA have different primary focuses, they can complement each other when used together.

Various approaches and tools complement RCM applications, such as:

- Risk analysis: Systematically assessing potential equipment and function failures, their consequences, and solutions.
- Design review based on failure modes (DRBFM): Primarily applying during design phases to identify potential failure modes, serving as a good reference for RCM analysis and planning.
- Bottleneck identification: Central to throughput management, discussed throughout the book. The process uncovers production constraints often caused by unreliable equipment.
- Downtime analysis: Investigating the patterns and causes of downtime to guide maintenance in reducing future occurrences (refer to Chapter 6).
- Drill-down process: An effective process for finding root causes in complex systems (refer to Chapter 6).

When applicable, applying these tools in conjunction with RCM can enhance effectiveness. Identifying the root causes of potential equipment failures related to reliability is crucial for effective maintenance planning. Another powerful tool for this purpose, fault tree analysis, will be discussed in the subsequent subsection 5.2.2.4.

5.2.2.4 Fault Tree Analysis

Fault tree analysis (FTA) is a reliability analysis method that delves into the root causes of system failures, aiding in risk reduction. It is particularly valuable for studying the relationships and probabilities of system and component failures within RCM.

FTA utilizes a map format, resembling an upside-down tree and employs Boolean logic gates to describe the relationships between elements at multiple levels. Gates, such as AND, OR, Priority AND, Exclusive AND, and Exclusive OR, have inputs and outputs with defined logical relationships. Figure 5.15a illustrates a typical format of FTA, while Figure 5.15b provides an example of the general processing module of a NASA computer platform [Siu et al. 2019].

FTA follows a deductive, top-down approach to identify the causes of specific failures in a complex system. It breaks down into the root causes of failure and their contributing factors with graphic visualization. In many cases, the probability of each failure can be estimated and incorporated into the FTA map.

Standardized across industries, FTA references include:

- Nuclear: NUREG–0492 [NRC 1981]
- Civil aerospace: SAE ARP4761 [SAE 1996]
- Military: MIL–HDBK–338B [DOD 1998]
- General industry: NASA FTA handbook [NASA 2002] and IEC 61025 [IEC 2006]

FTA has been successfully applied to various applications, including the design reliability of an electromechanical system [Jiang et al. 2019] and assessing fire risk in a gas processing plant [Hosseini et al. 2020]. While its applications in maintenance management are currently limited, this suggests opportunities for further development and research in this field.

Subsection Summary: This subsection examines the failure patterns associated with equipment aging, the RCM process, and its applications. It also discusses the applicability and potential of FTA in maintenance. All the reliability-related principles and approaches can enhance equipment maintenance and ultimately contribute to system excellence.

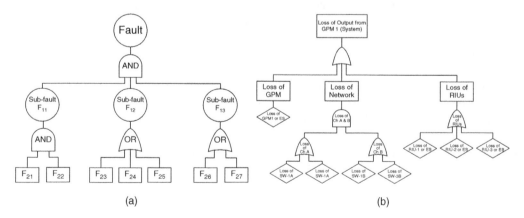

Figure 5.15 Examples of fault tree analysis.

5.3 Maintenance Performance Management

5.3.1 Key Aspects of Maintenance

5.3.1.1 Comparison of Maintenance Strategies

As discussed in Section 5.2, the three basic maintenance strategies are corrective, preventive, and predictive. In addition, TPM can be considered an all-inclusive strategy for maintenance, and RCM is viewed as a predictive strategy. Each strategy has its own characteristics, but management can jointly apply them in practice. Table 5.4 compares the technical features of these maintenance types.

This comparison serves as a comprehensive and informative reference for selecting maintenance strategies. Note that in the table, the evaluations of many items, such as cost and risk, are relative and may vary depending on the specific application.

From the comparison above, it can be concluded that corrective maintenance involves fixing problems without a plan, implying downtime. It is not the best approach for ensuring system performance, except for certain situations such as for unimportant or parallel functions.

Preventive maintenance, on the contrary, is a planned activity based on historical data and experience. The scheduled maintenance should occur during nonproduction periods to avoid operational interruptions. Being conservative, it can be performed earlier than strictly necessary.

Predictive maintenance is highly effective as it aligns with the actual needs. However, it requires suitable technology to sense the conditions and/or analyze reliability and risks, which require upfront work and investment and are contingent on technical feasibility and financial justification.

Table 5.4 Characteristics of basic maintenance strategies.

Strategy	Corrective (Reactive to failures)	Preventive (Schedule-based)	Predictive (Condition-based)	Predictive (Reliability-centered)
Objective	To fix problems	To prevent problems	To predict and prevent problems	To predict and prevent problems
Trigger	Problems, e.g., down, malfunction, etc.	Quantity, e.g., time, cycles, units, etc.	Signals, parameters, etc.	Historical data, reliability analysis, etc.
Activity	To respond to issues or malfunctions	To inspect and maintain	To monitor and diagnose	To analyze potential failures to predict
Timing	Unplanned	Scheduled, fixed	Real time, adaptive scheduling	Scheduled
Maintenance cost	Low	Low or medium	Medium	Medium
Technology and skillset	Low	Low	Medium	High
Risk of downtime	High	Low or medium	Low	Low
Total cost (including throughput)	High	Medium	Low	Low

Selecting appropriate maintenance strategies should be driven by the total maintenance cost, rather than solely on maintenance expenses. This key point will be further discussed later. Numerous studies provide insights into maintenance strategy selection and evaluation including Pariazar et al. [2008], Gandhare and Akarte [2012], Garcia-Teruel [2022], and Montero Jiménez et al. [2023].

5.3.1.2 Maintenance for Throughput

System throughput failures and other performance issues can happen either suddenly (e.g., breakdowns) or gradually (e.g., deteriorated quality attributes). Furthermore, the visibility of these issues varies; some are readily noticeable, while others require monitoring for detection.

To address these challenges, a system's performance should be accurately monitored in KPIs (discussed in Chapter 2) with predefined thresholds. As illustrated in Figure 5.16a, when an abrupt failure occurs, as shown in Figure 5.16b, immediate maintenance attention is imperative. If a monitored KPI falls and reaches its threshold, remedial actions can be initiated, as seen in Figure 5.16a.

Gradual performance deterioration warrants further discussion. Figure 5.17 illustrates how different maintenance strategies can affect manufacturing performance. The conclusion drawn from

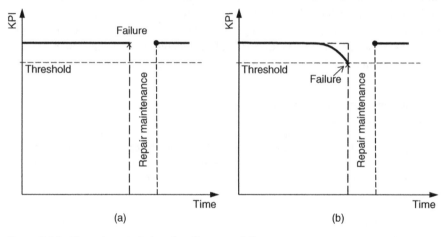

Figure 5.16 Time characteristics of performance failures.

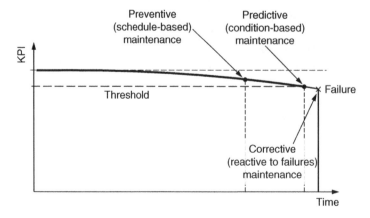

Figure 5.17 System performance and maintenance strategy timing comparison.

the figure aligns with the earlier discussion: corrective maintenance reacts to failures, which downtime or major issues already happen. On the contrary, both preventive and predictive maintenance may avoid downtime. However, preventive maintenance may be less cost-efficient than predictive maintenance, as the former acts somewhat too early.

For gradual failures or predicted issues, there is a time window for scheduling maintenance activities. Preventive actions can be conducted during scheduled breaks or nonproduction hours. In addition, when the downstream buffer contains more WIP units than average, it may allow maintenance to address certain types of issues without starving downstream systems. For the same reason, when practical and appropriate, production control may intentionally increase WIP units in the related buffers for maintenance activities to minimize their impact on operations. The opportunities to repair windows can be evaluated and realized based on maintenance prediction and buffer status.

The degradation monitoring of equipment performance can be realized through various attributes and variables such as torque, accuracy, pressure, temperature, sound, speed, and wear. The thresholds for performance degradation, which depend on specific equipment's functions and characteristics, require analytical work up-front. Moreover, the study of equipment performance degradation can serve as a valuable research topic for RCM and its applications.

In a large manufacturing system with similar/same equipment, two maintenance approaches can be employed: individual maintenance and group maintenance. Group maintenance involves repairing or replacing components in all similar machines when one experiences a failure or is predicted to fail soon. Given similar component reliability, group maintenance can better ensure system reliability. However, it is more costly and time-consuming compared to individual maintenance. Thus, selecting which approach depends on several factors, including the importance of system performance, estimated risks, and resource constraints.

5.3.1.3 Time Consideration in Maintenance

Moving from maintenance strategies to timing considerations, one concern is the frequency or interval of inspections, when real-time inspection is not available. Tailoring monitoring and inspection schedules to specific equipment performance can improve the effectiveness of both preventive and predictive maintenance. Figure 5.18 illustrates an example where the inspection frequency increases (or interval decreases) as the equipment is used more.

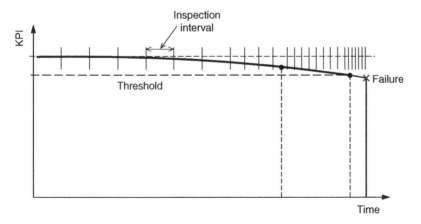

Figure 5.18 Varying inspection intervals maintenance.

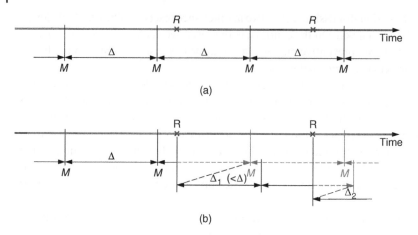

(a)

(b)

Figure 5.19   (a) Fixed scheduling and (b) adaptive scheduling of preventive maintenance.

As shown in the figure, when the KPI measurement approaches the predetermined threshold, the frequency is increased (or the interval is reduced, for instance, halved), because of the higher probability of failures. When the KPI value is very close to the threshold or the equipment/component nears the end of its lifespan, the inspection frequency should be further increased. This approach ensures that maintenance is performed precisely when needed, maximizing equipment lifespan while minimizing maintenance cost and risks.

Particularly for preventive maintenance, the time interval of maintenance tasks can be in two schedules. One schedule keeps fixed intervals, as shown in Figure 5.19a. This simple way does not consider the status, failures, and/or repair efforts of the equipment. In the figure, "M" stands for scheduled maintenance work, and "R" represents repairs of failures that occurred.

Alternatively, as demonstrated in Figure 5.19b, this approach involves adjusting the intervals based on the occurrence of failures for subsequent maintenance tasks. Using this approach, the current schedule is reset and starts over when a repair is conducted. The issue that occurred implies it is less reliable than expected or happened earlier than the scheduled maintenance. Therefore, the new interval (Δ_1) is set shorter than the original one (Δ) to avoid issues happening before the scheduled maintenance. This approach sounds more logical and beneficial to system performance excellence.

Subsection Summary: This subsection covers various facets of effective maintenance management, including strategy comparison and selection. It reviews the considerations for sudden and gradual failures, tailored monitoring and maintenance scheduling, and group maintenance for enhancing system performance.

5.3.2   Measuring Maintenance Quality

5.3.2.1   Key Aspects of Maintenance Quality

The effectiveness or quality of maintenance can be evaluated based on operation performance as it is its ultimate goal. The value of good maintenance becomes clear when downtimes and defective parts are at low levels and maintenance cost are reasonable. Even in these favorable situations, quantitatively measuring maintenance effectiveness for business operations remains important to identify further improvement opportunities.

The Association of German Engineers (VDI) developed the VDI 2887 Quality Management in Maintenance guideline and updated it in 2019 [VDI 2019]. According to VDI 2887, maintenance

Table 5.5 Types of maintenance audits.

Audit component	Product audit	Process audit	System audit
Object	Focus on completed maintenance tasks	Focus on maintenance process (order processing)	Focus on maintenance or quality management
Purpose	Follow up on quality characteristics	Ensure order processing compliance	Evaluate management system effectiveness
Criteria	Based on system functionality, order characteristics (e.g., time and cost compliance)	Based on resources, efficiency and effectiveness, workflows, etc.	Based on general requirements of standards (e.g., ISO 9001)
Documents	Notification/order, drawings, plant files, etc.	Orders, warranty, work plans, job description, training, etc.	Process documentation, organizational guidelines, etc.

quality is defined as "the degree to which the inherent characteristics of the maintenance, the functional, process, and occupational safety of a plant meet during its life cycle, considering and fulfilling all requirements of internal and external stakeholders."

The guideline addresses maintenance quality in four main aspects:

1. Quality planning in maintenance: This involves two key responsibilities: (1) evaluating all requirements for the maintenance processes, results, and systems and (2) planning to meet these requirements (Figure 5.10).
2. Quality control in maintenance: This aims to meet maintenance quality requirements by applying various principles and processes, including data, measurement, and analysis with a focus on maintenance activities.
3. Quality assurance in maintenance: Similar to quality control, this ensures long-term functionality through elements such as risk analysis, evaluation, documentation, and audits. Table 5.5 lists three audit types in maintenance practice, for assessing its effectiveness.
4. Quality improvement in maintenance: This follows the principles and processes of continuous improvement, which are discussed in more detail in Chapters 4 and 6.

A maintenance audit uses a combination of document and historical data reviews, interviews, and observations against predefined requirements and standards. Noncompliance findings should be documented and followed by corrections.

5.3.2.2 Maintenance Performance Measurement

Effective maintenance performance measurement is crucial for monitoring maintenance effectiveness and continuously improving maintenance activities. This uses KPIs related to both maintenance and system performance. The common maintenance KPIs are listed in Table 5.6, along with goal examples.

In addition to monitoring maintenance-specific KPIs, it is essential to consider system KPIs as maintenance plays a significant role in supporting system performance. Here are some common system KPIs related to maintenance effectiveness, listed in Table 5.7. Note that other factors also influence these system KPIs. Further discussion on these topics is provided in the next section 5.4.

Table 5.6 KPIs and corresponding goals of maintenance quality.

Maintenance KPI	Goal example
Safety (number of incidents)	<1 per month
Safety (severity, e.g., lost day)	<0.5 per quarter
Preventive rate (work hours)	>70%
Planned maintenance work	>95%
Wrench time (tool time)	>60%
Maintenance cost per unit	<$0.5
Maintenance overtime work	<10%
Spare part availability (in time)	>90%
Human errors	<2%
Skill training (hours/year)	>40

Table 5.7 System KPIs related to maintenance effectiveness.

System KPI	Goal example	Most related book subsection
MTTR	<10 minutes	5.1.2 (MTBF and MTTR)
MTBF	>20 hours	5.1.2 (MTBF and MTTR)
Availability	>97%	1.1.2 (Throughput Performance)
Cycle time	>98%	1.2.2 (Time Analysis)
OEE	>85%	2.2.2 (Introduction to OEE)
FTQ or TPY	>96%	4.3.2 (Quality Analysis)

These KPIs are good candidates for measuring maintenance performance. The selection and applications of maintenance KPIs and system KPIs involve judgment calls by management, including the setting of KPI targets. In addition, other KPIs, such as work order complete on-time rate, can be introduced. When choosing KPIs, key factors include:

- Related to organizational objectives
- Easy to understand
- Reliable data collection
- Responsive to changes

The general process and considerations for KPI selection are discussed in Section 2.4. Specifically for the relationship between maintenance and operations, a key is the quantitative correlation between the two types of KPIs. The strongly associated ones, for instance, MTTR and operational availability, can be good candidates.

5.3.2.3 Common Maintenance KPIs

In addition to the maintenance KPIs and goals listed in Table 5.6, maintenance effectiveness can be measured. One such metric is the planned maintenance percentage (PMP), which measures the proportion of planned maintenance tasks compared to all maintenance tasks over a period.

$$\text{PMP} = \frac{\text{Planned maintenance hours}}{\text{Total maintenance hours}}$$

For example, if total maintenance work took 500 hours, including 85 hours of unscheduled work, in a month, then:

$$\text{PMP} = \frac{500 - 85}{500} = 83\%$$

A high PMP indicates that most maintenance work is based on schedules with only a small portion (17% in the example) involving reactive maintenance. Conversely, a small PMP, for example, <60%, suggests a significant reactive maintenance effort and likely resulted from high system/equipment downtime.

A similar metric is the ratio of unscheduled downtime to scheduled downtime of a system. The ideal situation is when this ratio is zero, indicating no unscheduled downtime in a given period. The actual target can be case-dependent and subject to continuous improvement.

$$\text{Unscheduled downtime ratio} = \frac{\text{Unscheduled downtime}}{\text{Scheduled downtime}}$$

Another KPI is maintenance schedule compliance (MSC), which measures the number of completed work orders within a specific timeframe. MSC is calculated as follows:

$$\text{MSC} = \frac{\text{Number of completed maintenance orders}}{\text{Number of scheduled maintenance orders}}$$

MSC provides an overall picture of planned work completion and reflects the maintenance resources and demands.

Combining maintenance cost and throughput performance, a 2D chart can be created to display the status and progress over time, as shown in Figure 5.20. The horizontal axis represents maintenance efforts, in terms of cost or time. The vertical axis represents a system throughput KPI, for example, downtime or throughput rate. When possible, the downtime can be presented in a converted dollar amount.

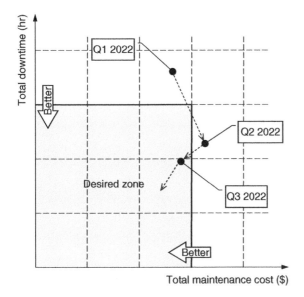

Figure 5.20 Maintenance cost-effectiveness tracking chart.

With maintenance cost-effect tracking, maintenance management and operational management can jointly determine the target levels for downtime (or other system KPI) and maintenance investment and expenditure. The desired target is represented as a zone in the figure rather than a fixed value.

Subsection Summary: Maintenance effectiveness can be evaluated from both internal (maintenance itself) and external (system performance) aspects. Analyzing the KPIs of both maintenance itself and the supported system allows focused improvement of maintenance practice to enhance system performance.

5.4 Consideration and Analysis in Maintenance

5.4.1 Risk and Effect Considerations

5.4.1.1 Risk-based Prioritization

To prioritize maintenance tasks, the risks of potential failures should be evaluated based on their probabilities and consequences. For example, if a conveyor motor failure has a very low probability but medium downtime impact on system throughput, it could be given a low priority.

Potential equipment failure probability can be rated at three levels: low, medium, and high. The rating can be based on historical data, such as MTBF, and reliability analysis. Similarly, the consequence severity of a potential equipment failure on system throughput can be assessed based on MTTR, repair costs, safety concerns, environmental issues, etc. The rating can be performed by a small team using a whiteboard based on team brainstorming and analysis.

For the major maintenance tasks on multiple potential failures, their probability and severity can be rated at the three levels. Figure 5.21 illustrates an example of a maintenance priority chart based on the evaluation of probability and consequence. In the figure, the distance from the origin to each task's location visually represents its relative priority or importance.

In the figure, potential failure ② has a high priority to be addressed because of its relatively high failure probability and consequence severity. The corresponding maintenance tasks should

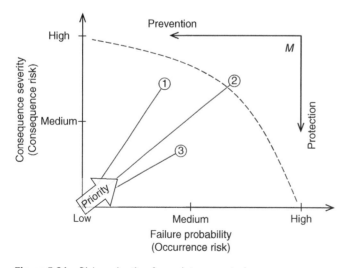

Figure 5.21 Risk evaluation for maintenance tasks.

be conducted first. The next maintenance target should be on potential failure ① as it is more important than ③.

Maintenance concepts, principles, and practice are active research. For example, some studies used computer simulation to find optimal maintenance plans [Alrabghi and Tiwari 2015]. Other studies sought to develop approaches and guidance rules for maintenance activities via modeling [Huang et al. 2018; Saez et al. 2022]. In their modeling, one or multiple KPIs, such as system operational availability, total maintenance cost, and maintenance efficiency, can be the objectives.

5.4.1.2 Recurrence Risk Considerations

When a piece of equipment fails, there are two scenarios for maintenance: repair (minor or major) and replacement. Each type involves varying levels of effort, time, and investment to restore the equipment's functionality. The relation between maintenance and the risk of failure reoccurring is a crucial consideration for a maintenance decision.

Repair restores functionality but may not significantly alter the aging status of the component/equipment. In some cases, the same failure can recur within a relatively short period. While replacement may be costly, it resets the risk of recurring failures to the original level. From a throughput standpoint over time, replacement can be preferred because it offers a low chance of recurring failures. Figure 5.22 compares the risks between minor repair, major repair, and replacement. If the failure risks can be converted to monetary terms, they can be directly compared with downtime losses for informed decision-making.

A quantitative assessment of failure risks in a system can be performed using reliable data such as equipment historical information and reliability data. However, in practice, accurately quantifying failure risks can be challenging due to uncertainties in reliability data, limited historical records, and varying operating conditions. Both the risk and cost-effectiveness of maintenance actions can be only roughly estimated.

Given the challenges of quantifying failure risks in complex systems, the criterion that combines the total maintenance cost and its impact on manufacturing performance becomes crucial. For instance, a good strategy is to conduct a minor repair, minimizing the interruption of ongoing operations. Then conducting a major repair or complete replacement is arranged during later non-production time. That strategy can benefit low impact on operation in the short term and low risk in the long run.

Subsection Summary: This subsection addresses how to evaluate maintenance task priorities through an assessment of the probability and severity of failure risks on manufacturing performance, as well as the recurrence risks after maintenance.

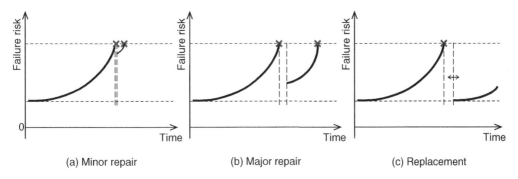

(a) Minor repair   (b) Major repair   (c) Replacement

Figure 5.22 Repair strategies and failure recurring risks.

5.4.2 Analysis of Maintenance Cost

5.4.2.1 Total Maintenance Cost

In addition to risk considerations, the costs of maintenance are another key factor. Maintenance directly contributes to a company's business operations. Typically, increasing maintenance efforts and investments, such as resources, workforce, spare parts, and technology, can yield benefits:

- Improved operational availability, resulting in lower throughput costs.
- Enhanced product quality, reducing quality costs due to rework and scrap (discussed in Chapter 4).

From a systems viewpoint, a total maintenance cost model provides a complete measure consisting of three types of cost:

$$\text{Cost}_\text{total} = \text{Cost}_\text{maintenance} + \text{Cost}_\text{throughput} + \text{Cost}_\text{quality}.$$

- Maintenance cost ($\text{Cost}_\text{maintenance}$): Include expenses related to labor, equipment, spare parts inventory, and overhead.
- Throughput cost ($\text{Cost}_\text{throughput}$): Measured by monetary items converted from throughput losses, including downtime, lost revenue, wages paid during downtime, and overtime work.
- Quality cost ($\text{Cost}_\text{quality}$): Encompasses rework and scrap resulting from defects and related quality activities.

These three types of cost are interrelated: For instance, better maintenance can result in less throughput loss. Figure 5.23 illustrates two examples. In the figure, the total cost curve can be viewed in the following three zones:

- Minimal maintenance zone (A): Minimal maintenance costs, but remarkably high throughput (downtime) and quality (defect) costs
- Best maintenance zone (C): Low throughput and quality costs, but high maintenance costs
- Lowest total maintenance cost zone (B): Minimal total maintenance cost

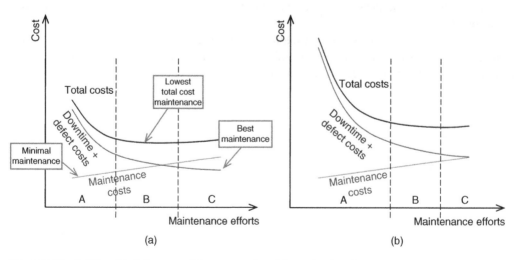

Figure 5.23 Relationship between maintenance cost and throughput costs.

The specific shape and location of these zones vary depending on the system. In most cases, management with a systems viewpoint can manage the total cost in the zone of low total cost. However, in some cases, management may focus more on maintenance cost, resulting in higher total cost due to increased throughput and quality costs.

Furthermore, maintenance cost fluctuates over time due to various factors, including equipment age and usage. Newly acquired equipment typically requires only routine checks and basic maintenance. Over time, as equipment usage increases, more frequent and extensive maintenance becomes necessary. One study found that the maintenance cost for a machine increased by 59% from the first year to the second year, by 12% from the second year to the third year, and by 18% from the third year to the fourth year [Hagström et al. 2020].

Overall, analyzing these cost items and factors provides valuable insights for maintenance management. The next subsection 5.4.2.2 will further explore cost considerations.

5.4.2.2 Maintenance Cost Justification

Without detailed cost analysis, companies may remain unaware of whether their maintenance practices fall within the "lowest total cost" zone (Figure 5.23). A reason could be the lack of reliable data justifying adjustments in changes of maintenance investment and practice.

For such an analysis, the first step is to establish current cost levels as a baseline, allowing for the observation of changes in throughput and quality as maintenance practices evolve. Such data enables the estimation of the new total maintenance cost and determining whether increasing or decreasing maintenance is warranted, along with the next steps. One challenge lies in gathering sufficient data and keeping it current when introducing new processes and technologies.

Two effective metrics are the maintenance cost per product unit and total maintenance cost per unit:

$$\text{Unit cost}_{\text{maintenance}} = \frac{\text{Cost}_{\text{maintenance}}}{\text{Units produced}}$$

$$\text{Unit cost}_{\text{total}} = \frac{\text{Cost}_{\text{total}}}{\text{Units produced}}$$

As discussed previously, the total maintenance cost ($\text{Cost}_{\text{total}}$) includes three costs: $\text{Cost}_{\text{maintenance}}$, $\text{Cost}_{\text{throughput}}$, and $\text{Cost}_{\text{quality}}$.

With such data over time, the relationship between the two types of unit costs can be analyzed. Figure 5.24 shows an example. In this case, increasing maintenance cost leads to improved total unit costs or throughput.

In such an analysis, pay more attention to the overall trends, especially to the total cost, rather than the details as they can be influenced by other factors.

Importantly, maintenance is typically a major, but may not be the sole contributor to throughput. It is also possible that maintenance is not a major contributor to throughput at certain times. In some contexts, the costs of quality issues can exceed maintenance cost. Despite this, a total maintenance cost curve remains valuable for guiding maintenance planning and continuous improvement. For maintenance improvement projects, the economic evaluation approaches introduced in subsection 4.2.3 of Chapter 4 can be valued tools.

Subsection Summary: Cost analysis, particularly when based on total maintenance cost, can provide reliable guidance and meaningful justification for maintenance management. This is essential for achieving effective maintenance and optimal operations.

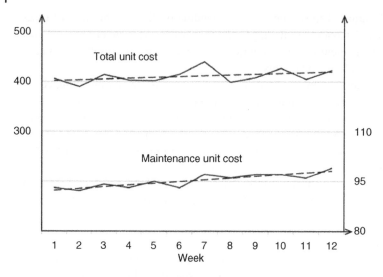

Figure 5.24 Trend monitoring of maintenance unit cost and total unit cost.

5.4.3 Additional Considerations in Maintenance

5.4.3.1 Impact Assessment on OEE

As discussed previously, maintenance tasks should be prioritized primarily based on their impact on system performance. The impact can be assessed using OEE (discussed in Chapter 2). The impact on the three OEE elements, namely, operational availability, speed performance, and product quality, can be individually assessed and rated against maintenance impact.

Table 5.8 displays an example of the maintenance impact ratings for OEE elements in a mass production environment. In this example, the impact on operational availability is considered the most important. Thus, the impact rating on operational availability (R_A) is on a scale of 10, while those on the other two elements (R_P and R_Q) are rated on a scale of 5. Note that this is an illustrative example; actual rating scales and details should be based on the specific application.

Table 5.8 Impact rates of maintenance tasks on system OEE.

Availability (A)	R_A	Performance (P)	R_P	Quality (Q)	R_Q
None	1	None	1	None	1
Downtime < 5 seconds	2	Reduction < 0.5 minute	2	Defect < 0.5%	2
Downtime < 15 seconds	3	Reduction < 1 minute	3	Defect < 1%	3
Downtime < 30 seconds	4	Reduction < 1.5 minutes	4	Defect < 2%	4
Downtime < 45 seconds	5	Reduction > 1.5 minutes	5	Defect > 2%	5
Downtime < 1.5 minutes	6				
Downtime < 3 minutes	7				
Downtime < 5 minutes	8				
Downtime < 10 minutes	9				
Downtime > 10 minutes	10				

The impact of maintenance tasks on system operations can be assessed by the product of the three individual scores, a priority identification (PI) number:

$$PI = R_A \times R_P \times R_Q$$

For example, if the impact scores of a maintenance task are 6, 4, and 3 for the three elements based on Table 5.8, respectively, the PI number for this maintenance task is:

$$PI = 6 \times 4 \times 3 = 72.$$

It is worth emphasizing that the PI numbers assigned to maintenance tasks represent their relative priority levels, without any meaning to the tasks.

5.4.3.2 Criticality with Additional Factors

To further refine maintenance prioritization beyond OEE impact, criticality evaluations of other key factors, such as safety and equipment value, can be conducted. These factors contribute to prioritizing maintenance when addressing potential issues.

Considering all important aspects, maintenance management can summarize the tasks into four priority levels: A–D. Table 5.9 presents this priority categorization in a high-value production context, outlining issue characteristics and corresponding maintenance task priorities. While this approach simplifies planning work, its quantitative and subjective nature has limitations on the reliability and justifications.

In addition, maintenance planners must consider schedule compliance. Preventive maintenance sometimes becomes overdue due to resource limitations. Operating equipment beyond its maintenance schedule increases risks of inferior functionality, downtime, and/or quality issues. In such situations, priority should be given to mitigate the risks of potential failures.

Furthermore, certain manufacturing operations, like batch processes, have varying equipment utilization rates. Maintenance may differ accordingly, with high-use equipment requiring more frequent attention than low-utilization equipment. For instance, a machine used 80% of the time requires more maintenance than one used 30% of the time. Therefore, factoring utilization rates into prioritization optimizes resource allocation, preventing under-maintenance of high-use equipment and over-maintenance of low-use equipment.

5.4.3.3 Maintenance Advancement

The discussion thus far has primarily focused on manual maintenance. Now, let us explore potential advancements in maintenance. One avenue is maintenance automation and self-maintenance. This involves automatic maintenance execution based on equipment status, maintenance records, resources, and schedules, using appropriate technologies.

Table 5.9 Maintenance task priority for different types of issues.

Priority	Issue type	Recommended maintenance priority
A	Server, related to safety or regulation	Immediately fix them, regardless of downtime duration
B	Critical to operation or quality	Immediately fix them, possibly temporary to minimize downtime or uphold quality, with follow-up actions
C	Risky and common, but not urgent	Fix them during the next break time
D	Nominal, supportive	Fix them after higher priorities are completed

Maintenance automation is particularly relevant in predictive maintenance. Equipment conditions, known through various sensors and analyses, enable specific maintenance tasks to be conducted with minimal impacts on production operations at the lowest total cost.

Existing studies explore maintenance automation feasibility in various sectors such as road maintenance [Eskandari Torbaghan et al. 2020], small hydropower plants [Selak et al. 2015], pyroprocess equipment [Han et al. 2019], and product features recognition [Yepez Herrera 2019]. Another study reviewed advancements in combining machine learning, infrared thermography, and augmented reality for industrial equipment maintenance [Venegas et al. 2022].

Challenges in achieving maintenance automation include acquiring technology and tools for performing physical maintenance tasks, developing automated support systems for real-time decision-making and actions, and ensuring reliable data to support maintenance decisions. Notably, economic justification through break-even and cost-effectiveness analyses plays a pivotal role. While promising, maintenance automation remains a long-term goal for many manufacturers.

Subsection Summary: Maintenance prioritization requires considering the impact on throughput, as presented by OEE, criticality, safety, equipment value, and scheduling compliance. A comprehensive approach to all factors may simplify maintenance planning. In addition, maintenance automation presents a promising future direction.

Chapter Summary

5.1. Maintenance Principles

1. Reliability changes over time and can be expressed as the probability that a system will perform its intended functions.
2. Mechanical components and products may have a bathtub-like failure curve with three distinct phases. There are other failure rate curves besides the bathtub curve.
3. MTBF (mean time between failures) measures reliability. MTTR (mean time to repair) measures maintenance efficiency.
4. Maintenance management is an integral component of plant management, working alongside operation management, quality management, and other supporting functions.
5. Maintenance, impacting both availability and quality, contributes to system throughput.

5.2. Equipment Maintenance Strategies

6. Maintenance planning integrates with asset lifecycle investment management for optimal system performance, aligning with overall business objectives.
7. Corrective, preventive, and predictive maintenance are common practices, having different risks and impacts on system performance.
8. Total productive maintenance (TPM) is a comprehensive approach that involves all manufacturing personnel, operations, and methods. A key component of TPM is AM (autonomous maintenance).
9. Reliability-lefted maintenance (RCM) incorporates all three maintenance types, emphasizing prediction and prevention, focusing on system throughput.
10. Fault tree analysis (FTA), a reliability analysis method, identifies causes of system failures, employing deductive reasoning through a graphical map.

5.3. Maintenance Performance Management

11. The choice of maintenance strategy depends on total maintenance cost, failure response needs, and preventing downtime through timely interventions.
12. Differentiating between gradual and sudden performance issues, maintenance strategies vary to react to failures or prevent downtime, emphasizing the importance of timely interventions.
13. Cost considerations in maintenance involve choosing between group and individual approaches, influenced by operational importance and resource constraints.
14. Maintenance quality, as defined by VDI 2887, encompasses planning, control, assurance, and improvement, with audits playing a vital role in assessment.
15. Maintenance performance can be measured by various KPIs, such as safety, preventive rate, maintenance cost, but more importantly system KPIs such as an MTTR, MTBF, and OEE.

5.4. Consideration and Analysis in Maintenance

16. Evaluation of risks and impacts is important for repair and replacement decisions. Challenges include quantifying failure risks, addressing total maintenance cost.
17. Maintenance tasks should be prioritized based on the severity and risk of failures and associated costs.
18. Total maintenance cost considerations for maintenance planning include maintenance, downtime, and quality/defect costs.
19. Maintenance KPIs include PMP (planned maintenance percentage), MSC (maintenance schedule compliance), ratio of unscheduled downtime to scheduled downtime, and total maintenance cost.
20. Maintenance automation and self-maintenance through technologies like AI and advanced sensors is an advancement direction.

Exercise and Review

Exercises

5.1 A robot has a failure probability of 0.0002 per hour. Calculate its reliability for eight hours of work. If the failure probability is reduced to 0.0001 per hour, what would be the improved reliability over eight work hours?

5.2 Over a three-month production period, the four most frequent failure modes (A, B, C, and D) occurred 10, 13, 9, and 8 times, respectively. The average downtime for these modes was 8, 3, 12, and 7 minutes, respectively. If the consequence rating is considered the same for all modes, which failure mode should be prioritized for maintenance?

5.3 An assembly line encountered 92 failures over a 3-month production period, amounting to 900 working hours. Determine the MTBF for this line.

5.4 The maintenance department spent 1050 minutes rectifying 92 failures on an assembly line over a 3-month period. What is the MTTR for this line?

5.5 If the MTBF of a production line is 15.5 hours for a given period, what would be the corresponding failure rate?

Table 5.10 For Exercise 5.9.

Issue	A	P	Q
1	Downtime = 4 minutes	Slow by 0.6 minutes	None
2	None	Slow by 2.5 minutes	Defect = 1.9%
3	Downtime = 8 minutes	None	Defect = 0.3%

5.6 If the MTBF and MTTR of a production line are 15.5 and 0.25 hours, respectively, for a given period, calculate the availability of this production line.

5.7 During a month's production, unplanned maintenance for downtime accounted for 125 out of the total 600 maintenance hours. What would be the PMP and the unscheduled downtime ratio?

5.8 A maintenance department incurred expenses of $85k and $100k on various maintenance tasks over two consecutive months. The estimated costs for throughput and quality for these two months were $250k and $230k, respectively. Calculate the total maintenance cost for these two months.

5.9 A maintenance department adheres to the three elements of OEE to guide its maintenance activities. During a shift, three fundamental issues are reported, as listed in Table 5.10. Due to resource constraints, the maintenance department can immediately address only two of these issues. Which two should be prioritized for immediate attention? (Refer to Table 5.8 for the ratings).

Review Questions

(The chapter covers these topics. For further discussion, it is recommended to seek additional information and examples. Diverse perspectives are encouraged.)

5.1 Analyze the relationship between the failure rate and the reliability of a machine. Provide an example.

5.2 Explore the concept and potential application of failure severity evaluation, considering its three factors: frequency, duration, and consequence.

5.3 Do you agree that the reliability bathtub curve is primarily applicable to mechanical components? Please explain.

5.4 Explain the meanings of MTBF and MTTR with an example.

5.5 Provide an example and illustrate how the equipment's reliability can change over time.

5.6 Investigate the roles of maintenance management and its relationship with operational management and quality management within plant management, providing an example.

5.7 Review the three zones of the relationship between maintenance and operational availability (refer to Figure 5.7) and discuss how to determine the appropriate level of maintenance based on this relationship.

5.8 Discuss how effective maintenance can enhance system throughput by improving quality, providing an example.

5.9 Explain the distinctive characteristics of corrective, preventive, and predictive maintenance strategies.

5.10 Describe the differences between conventional maintenance and TPM.

5.11 Discuss two reliability-time patterns, as shown in Figure 5.13, providing examples.

5.12 Compare the concepts of throughput-focused maintenance and reliability-lefted maintenance, highlighting both similarities and differences.

5.13 Discuss the role of risk evaluation in maintenance planning, providing an example.

5.14 Why might there be a need for more frequent inspections during periods of performance degradation?

5.15 Define the quality or effectiveness of maintenance and explain your definition.

5.16 Review two KPIs that are used to measure maintenance performance.

5.17 Discuss how to apply PMP, the unscheduled downtime ratio, and MSC with examples.

5.18 Review how to implement a total cost (in terms of unit cost) approach in maintenance practices.

5.19 Comment on the use of system OEE as guidance in maintenance planning.

5.20 Discuss the feasibility of automating maintenance and the challenges associated with it.

References

Ahmad, N., Hossen, J. and Ali, S.M. 2018. Improvement of overall equipment efficiency of ring frame through total productive maintenance: a textile case. The International Journal of Advanced Manufacturing Technology, 94, pp. 239–256.

Alrabghi, A. and Tiwari, A. 2015. State of the art in simulation-based optimisation for maintenance systems. Computers and Industrial Engineering, 82, pp. 167–182.

Antosz, K. and Ratnayake, R.C. 2019. Spare parts' criticality assessment and prioritization for enhancing manufacturing systems' availability and reliability. Journal of Manufacturing Systems, 50, pp. 212–225.

Bouslah, B., Gharbi, A., Pellerin, R. 2018. Joint production, quality and maintenance control of a two-machine line subject to operation-dependent and quality dependent failures. International Journal of Production Economics, 195, pp. 210–226.

DOD 1998. Military Handbook – Electronic Reliability Design Handbook, MIL-HDBK-338B, Defense Quality and Standardization Office, Fort Belvoir, Virginia: United States Department of Defense. https://www.navsea.navy.mil/Portals/103/Documents/NSWC_Crane/SD-18/Test%20Methods/MILHDBK338B.pdf. Accessed July 2020.

DOE 2010. Operations and Maintenance Best Practices - A Guide to Achieving Operational Efficiency, Release 3.0, Pacific Northwest National Laboratory, U.S. Department of Energy, https://www.energy.gov/sites/prod/files/2013/10/f3/omguide_complete.pdf. Accessed July 2022.

Eskandari Torbaghan, M., Kaddouh, B., Abdellatif, M., Metje, N., Liu, J., Jackson, R., Rogers, C., Chapman, D., Fuentes, R., Miodownik, M. and Richardson, R. 2020. Application of Robotic and Autonomous Systems for Road Defect Detection and Repair-A Position Paper on Future Road Asset Management.

Gandhare, B.S. and Akarte, M., 2012. Maintenance Strategy Selection. In: Ninth AIMS International Conference on Management, pp. 1330–1336. Maharashtra, India: AIMS International.

Gaonkar, A., Patil, R. B., Kyeong, S., Das, D. and Pecht, M. G. 2021. An assessment of validity of the bathtub model hazard rate trends in electronics. IEEE Access, 9, pp. 10282–10290.

Garcia-Teruel, A., Rinaldi, G., Thies, P.R., Johanning, L. and Jeffrey, H., 2022. Life cycle assessment of floating offshore wind farms: an evaluation of operation and maintenance. Applied Energy, 307, p. 118067.

Hagström, M.H., Gandhi, K., Bergsjö, D. and Skoogh, A., 2020. Evaluating the effectiveness of machine acquisitions and design by the impact on maintenance cost – a case study. IFAC-PapersOnLine, 53(3), pp. 25–30.

Han, J., Ryu, D., Kim, D., Lee, J., Yu, S. and Shin, M., 2019. A conceptual framework for equipment maintenance automation under a pyroprocessing automation framework. Science and Technology of Nuclear Installations, 2019.

Hosseini, N., Givehchi, S. and Maknoon, R. 2020. Cost-based fire risk assessment in natural gas industry by means of fuzzy FTA and ETA. Journal of Loss Prevention in the Process Industries, Volucella, 63, p. 104025. https://doi.org/10.1016/j.jlp.2019.104025.

Huang, J., Chang, Q., Zou, J., Arinez, J. and Xiao, G. 2018. Real-time control of maintenance on deteriorating manufacturing system. In 2018 IEEE 14th International Conference on Automation Science and Engineering (CASE), pp. 211–216. IEEE.

IEC 2006. IEC 61025 Fault Tree Analysis (FTA), Geneva, Switzerland: International Electrotechnical Commission.

IEC 2009. Dependability Management - Part 3-11: Application Guide – Reliability Centred Maintenance, Geneva, Switzerland: International Electrotechnical Commission.

ISO 2014. ISO 55001:2014 Asset Management, Geneva, Switzerland.

Jiang, R. 2013. A new bathtub curve model with a finite support. Reliability Engineering & System Safety, 119, pp. 44–51.

Jiang, Y., Gao, J.M., Sun, G., Wang, R.X., Zhang, P.F., Chen, K. and Ma, D.Y. 2019. Fault Correlation Analysis-based Framework for Reliability Deployment of Electromechanical System. In The 2nd International Workshop on Materials Science and Mechanical Engineering, 26–28 October 2018, Qingdao, China. IOP Conference Series: Materials Science and Engineering, 504, p. 012113. https://doi.org/10.1088/1757-899X/504/1/012113.

Montero Jiménez, J.J., Vingerhoeds, R., Grabot, B. and Schwartz, S. 2023. An ontology model for maintenance strategy selection and assessment. Journal of Intelligent Manufacturing, 34(3), pp. 1369–1387.

NASA 2002. Fault Tree Handbook with Aerospace Applications. Version 1.1, Washington, DC: National Aeronautics and Space Administration.

Nowlan F.S. and Heap H.F. 1978. Reliability Lefted Maintenance, San Francisco, CA: United Airlines Publications.

NRC 1981. Fault Tree Handbook (NUREG–0492). https://www.nrc.gov/reading-rm/doc-collections/nuregs/staff/sr0492/index.html. Accessed November 2022.

Pariazar, M., Shahrabi, J., Zaeri, M.S. and Parhizi, S. 2008. A combined approach for maintenance strategy selection. Journal of Applied Sciences, 8(23), pp. 4321–4329.

Roesch, W.J. 2012. Using a new bathtub curve to correlate quality and reliability. Microelectronics Reliability, 52(12), pp. 2864–2869.

SAE 1996. ARP4761 Guidelines and Methods for Conducting the Safety Assessment Process on Civil Airborne Systems and Equipment, Warrendale, PA: SAE International.

SAE 2009. Evaluation Criteria for Reliability-Lefted Maintenance (RCM) Processes JA1011_200908, Warrendale, PA: SAE International. https://doi.org/10.4271/JA1011_200908.

Saez, M., Barton, K., Maturana, F. and Tilbury, D.M. 2022. Modeling framework to support decision making and control of manufacturing systems considering the relationship between productivity, reliability, quality, and energy consumption. Journal of Manufacturing Systems, 62, pp. 925–938.

Sala, R., Bertoni, M., Pirola, F. and Pezzotta, G. 2021. Data-based decision-making in maintenance service delivery: the D3M framework. Journal of Manufacturing Technology Management, 32(9), pp. 122–141.

Sandberg, U., Ylipää, T., Skoogh, A., Isacsson, M., Stieger, J., Wall, H., Andersson, M., Johansson, H., Nilsson, I., Agardtsson, J. and Vikström, S. 2014. Working with forces promoting or hindering implementation of strategies for maintenance – experiences from Swedish industry, In Proceedings of Swedish Production Symposium, Gothenburg, pp. 16–18. September 2014.

Selak, L., Škulj, G., Sluga, A. and Butala, P. 2015. Assessing feasibility of operations and maintenance automation – a case of small hydropower plants. Procedia Cirp, 37, pp. 164–169.

Siddiqui, A. and Ben-Daya, M. 2009. Chapter 16 Reliability Lefted Maintenance. In: Handbook of Maintenance Management and Engineering, pp. 397–415. ISBN: 978-1848824713. London: Springer.

Siu, K.Y., Herencia-Zapana, H., Manolios, P., Noorman, M. and Haadsma, R. 2019. Safe and Optimal Techniques Enabling Recovery, Integrity, and Assurance (No. NASA/CR-2019-220283), Hampton, Virginia: National Aeronautics and Space Administration, Langley Research Left.

Tang, H. 2022. Quality Planning and Assurance – Principles, Approaches, and Methods for Product and Service Development, ISBN-13: 978-1119819271, Hoboken, NJ: Wiley.

Tóth, Z.E. and Jónás, T. 2014. Typifying empirical failure rate time series: a case study on consumer electronic products. In Proceedings of International Work-conference on Time Series (ITISE), pp. 396–407.

VDI 2019. VDI 2887 Qualitätsmanagement der Instandhaltung: Quality management in maintenance, Verein Deutscher Ingenieure, Beuth Verlag GmbH. December 2019.

Venegas, P., Ivorra, E., Ortega, M. and Sáez de Ocáriz, I. 2022. Towards the automation of infrared thermography inspections for industrial maintenance applications. Sensors, 22(2), p. 613.

Yepez Herrera, P. 2019. An intelligent framework for automatic maintenance plan generation based-on product features recognition (Master's thesis, University of Alberta). https://era.library.ualberta.ca/items/1bb73a5c-1861-41dd-b2bd-05bc13d75968. Accessed April 2023.

6

Throughput Enhancement Methodology

6.1 Approaches for Solving and Improving

6.1.1 Overview of Problem-Solving and Continuous Improvement

Problem-solving (PS) has long been a part of human activity. George Pólya's 1945 work "*How to Solve It*" [Pólya 1945] was one of the first methodical descriptions of PS processes. Continuous improvement (CI) also has a long history. W. Edwards Deming introduced it as a systematic approach to industry in the 1950s. Both have been applied across various fields and disciplines.

6.1.1.1 Comparison between PS and CI

PS and CI are structured processes and follow a similar flow with distinct focuses, as depicted in Figure 6.1. Their different steps and actions can be classified into the following three fundamental phases:

1. Problem understanding and identification: In this initial phase, the goal is to quantitatively describe the existing problem or improvement opportunity and establish a project plan. This phase should lead to objectives or targets for a PS or CI project. The difference between PS and CI lies in addressing problems for a solution versus identifying opportunities for improvement, respectively.
2. Root cause and influence factor analysis: In this diagnostic phase, data analysis is essential for identifying root causes in PS or the primary influence factors in CI, as illustrated in the figure to highlight a key distinction. Various analytical techniques and methods can be employed in this phase.
3. Solution suggestion and implementation: This phase focuses on developing solutions based on the root cause/factor analysis, assessing these solutions, and prioritizing implementation. It is equally critical for both PS and CI projects. After implementation, PS recovers from the problem; CI makes the state even better, as shown in the figure.

Because of the similarity in PS and CI processes, they may be considered identical, and the terms PS and CI sometimes are used interchangeably, even though there are some key differences between the two.

Subsequently, the mindset applied in these processes can be discussed. Traditionally, people think "if it ain't broke, don't fix it" [Martin n.d.], meaning there is no need to change something that works well. This mindset works for PS but not for CI. The CI mindset seeks ways to improve status and encourages evaluating the current state of work and identifying areas for improvement. From this perspective, PS is *reactive* to an existing problem, while CI is *proactive* in improving the status.

Manufacturing System Throughput Excellence: Analysis, Improvement, and Design, First Edition. Herman Tang.
© 2024 John Wiley & Sons, Inc. Published 2024 by John Wiley & Sons, Inc.
Companion website: www.wiley.com/go/Tang/ManufacturingSystem

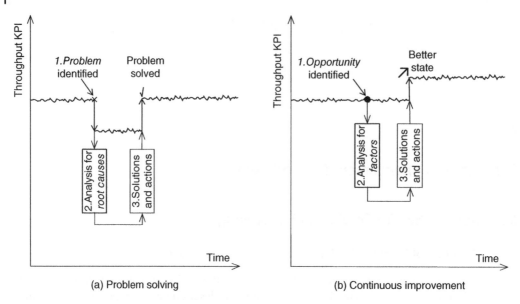

Figure 6.1 Comparison of problem-solving and continuous improvement.

In addition, adopting them as guiding principles and integrating them into routine operational management is essential and beneficial. By adopting the right mindset and principles, we can leverage both approaches to maintain satisfactory performance and continually make progress.

6.1.1.2 CI Targeting

In addition to the above discussion on the similarities and differences between PS and CI, setting targets is another uniqueness of CI. Unlike PS where a specific issue demands attention, identifying improvement opportunities in situations where everything functions well can be challenging.

A main way to choose targets for CI projects is self-assessment to identify a gap to a better state. Selected KPIs, such as TR and OEE, can be used to define both the current and expected states. As emphasized previously, a throughput enhancement target should be the output of the entire system, may or may not be the output of a subsystem or an operation area.

Benchmarking on other operations or organizations is an effective way to identify targets for CI. This involves understanding and comparing the current operational performances with those of best-in-class operations. Within a maintenance context, KPIs such as planned maintenance percentage (PMP) and system KPIs (as listed in Table 5.7 in Chapter 5) can aid in target selection. For instance, if the current PMP is 65% and an internal sister plant has achieved 70%, then the CI projects can aim for a target of 70%. Achieving a substantial target may necessitate the implementation of multiple incremental CI projects.

When a system's throughput is measured by a composite indicator, such as OEE, benchmarking studies can be conducted too (subsection 2.2.3). As manufacturing processes and situations are not the same, it is crucial to understand the differences when comparing other operations' OEE for improvement. Some system OEE levels in different industries are mentioned in Chapter 2.

6.1.1.3 Reasoning in PS and CI

In both PS and CI processes, several types of reasoning (inductive, deductive) can be applied. An understanding of these reasoning processes can enhance the effectiveness of both PS and CI processes. Figure 6.2 shows a diagram of how induction and deduction work.

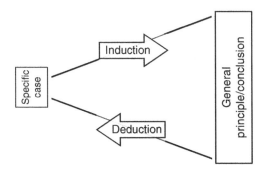

Figure 6.2 Inductive and deductive reasoning processes.

Deductive reasoning applies general principles to specific situations to reach a logical conclusion. For example, if all machines at the design speed produce 100 units/hour, machine A at optimal speed also produces 100 units/hour. Deductive reasoning is a common process. In throughput enhancement, projects normally employ established methods and approaches, such as reliability analysis. These general methods and approaches help identify the causes of problems in specific situations.

Inductive reasoning starts with specific observations and leads to a general conclusion or a new understanding. For instance, at optimal speed, machine A produces 100 units/hour, machine B produces 110 units/hour, and machine C produces 95 units/hour. Then, one may conclude that the production output of a machine depends on its capacity and performance.

Inductive reasoning can be applied in throughput enhancement to analyze the system performance using various KPIs, such as TR and TR_{sa}. As discussed earlier, it has been concluded that TR_{sa} accurately represents system performance as a useful tool for bottleneck identification, contributing to a new understanding. Even better, a new way of identifying and explaining bottlenecks can be generated. Within efforts in manufacturing excellence, new understanding and methodologies have been continually discovered.

In PS, it is often tempting to focus on specific problems and for quick solutions. Yet, many similar problems with common underlying root causes can be synthesized for general understanding and solutions, which can be more innovative and value-added, contributing to the professional fields. Figure 6.3 visually depicts the overall flow of this approach.

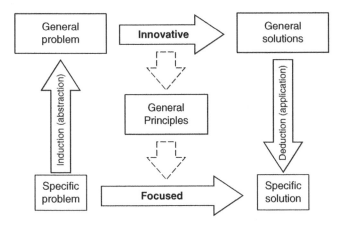

Figure 6.3 Focused and innovative problem-solving processes.

One well-known method for identifying general solutions is TRIZ (theory of inventive problem-solving, translated from Russian). There are many TRIZ applications and reviews in published articles [Spreafico and Russo 2016; Ekmekci and Nebati 2019; Sojka and Lepšík 2020].

Manufacturing professionals should remain open to adopting innovative PS approaches. By successfully implementing a generalized PS process like TRIZ, manufacturing professionals can not only find solutions to specific problems but also be capable of effectively addressing various problems.

Subsection Summary: This subsection explores PS and CI processes, comparing their similarities and differences, particularly on the CI mindset and target setting. It also examines reasoning processes and open-mindedness in and adopting new approaches.

6.1.2 Common Approaches

There are many approaches for PS and CI. This subsection offers a high-level overview of four widely used methods: DMAIC, 8D, A3, and PDCA.

6.1.2.1 DMAIC

As a structured data-driven approach, the DMAIC approach follows five steps: define, measure, analyze, improve, and control (DMAIC), as shown in Figure 6.4. These following five steps can be grouped into three phases: 1. Problem/Opportunity, 2. Root-causes/Factors, and 3. Solutions/Implementation, as discussed earlier.

1. The first step is to "*Define*" the problem and objective of a throughput enhancement project. The objective should follow SMART criteria: Specific, Measurable, Achievable, Relevant, and Time-bound. For instance, "Improve the TR of manufacturing system X by 0.3 JPH within four weeks." As mentioned, the objective should be assessed for its influence on the entire system.
2. The second step is to "*Measure*" the current status of the production throughput identified in the first step, including the target system, its upstream and downstream processes, and associated buffers. Relevant data, such as quality defects, should also be collected.

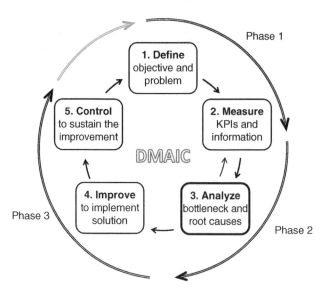

Figure 6.4 DMAIC five-step process flow and three phases.

3. The third step is to "*Analyze*" the data collected and brainstorm to identify bottlenecks (addressed in Chapter 3) and root causes. For bottleneck identification, the Measure and Analyze steps may need to be repeated. To find root causes, analysis may be conducted for multiple contributing factors, such as equipment reliability, process flows, specific tasks, or functions, etc.

4. The next step is to "*Improve*" the situation, which often requires additional resources and can be technically challenging. Following the implementation of solution actions, the throughput level should be re-evaluated by collecting and analyzing new data. Note that temporary solutions may be necessary in some cases, although they may not be suitable for the long run.

5. The last step is to "*Control*" the achieved new level and sustain the system on its new course. The "Control" approaches vary depending on the cases such as updated parameters, process settings, and standards. At this step, a project summary with lessons learned should be documented. As the final step in a project cycle, new ideas or potential problems for the next cycle may emerge.

Several examples highlight the application of DMAIC in throughput improvement. Hakala [2018] studied the layout and storage capacity of electrical manufacturing, Daniyan et al. [2023] examined the overall output of rail manufacturing processes, and Hardy et al. [2021] focused on finishing processes in a high-volume panel lamination line.

6.1.2.2 8D and A3

The 8D approach is a team-focused structured process with standardized eight disciplines. Like DMAIC, the 8D can be grouped into the three phases discussed earlier. The core elements of the 8D form the 5D approach, as shown in Figure 6.5. The 5D is more suitable for relatively small or simple projects.

D1: Form a cross-functional team with management support.
D2: Define and describe a problem that is specific and measurable, e.g., KPI and target.
D3: Contain the problem and keep production running with an interim plan.
D4: Identify the bottleneck and root causes using various tools.
D5: Develop a permanent corrective solution, considering effect, risk, cost, time, etc.
D6: Implement the corrective plan, verify results, and monitor the effect for a period.
D7: Sustain the improvement with certain measures.
D8: Close the project and recognize the project team.

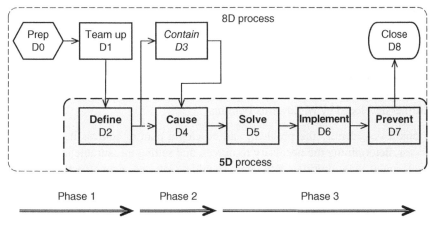

Figure 6.5 8D and 5D process flows and three phases.

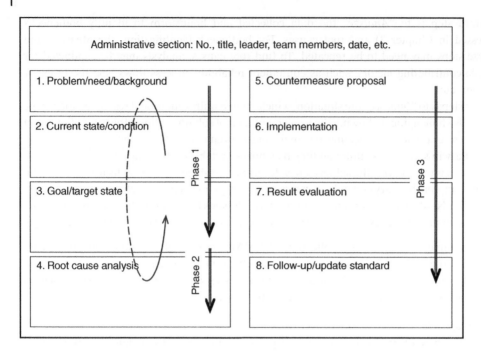

Figure 6.6 A3 problem-solving process/template and three phases.

An additional step, called D0, can be added to the 8D approach for preparation and setup. There are numerous studies and applications of the 8D approach, e.g., George et al. [2021].

Similarly, another widely used approach is called A3, which is named after the paper size on which it is typically documented in a one-page format. Introduced by Toyota [Sobek and Jimmerson 2004], this approach provides concise information about a PS process on a single page, as shown in Figure 6.6.

The A3 approach typically includes eight items to be addressed during a PS process. One advantage of the A3 approach is the effective communication of all information on one page. The A3 shares similarities with the 8D but has some minor differences. Widely applied in industries, published case studies of the A3 are less common compared to the 8D. For instance, Santos Filho and Simão [2022] provide an example of a published A3 application focused on reducing machine setup.

6.1.2.3 PDCA

Another popular approach is called PDCA (sometimes referred to as PDSA), which stands for Plan, Do, Check (or Study), and Act. The four-step approach is a cyclical framework for PS and CI. This cyclical framework has been widely applied since it was originally proposed by Walter A. Shewhart in the 1920s and further developed by W. Edwards Deming.

- Plan: Establish objectives, identify the problem, and develop a plan to address it, including defining the process, determining the necessary resources, and setting measurable targets. For throughput enhancement, an appropriate KPI target should be selected as an objective and for problem identification.
- Do: try out the plan on a small scale or in a controlled environment, for example, a pilot project or a test run. Particularly for throughput PS and CI, focus on bottlenecks and try different ways if applicable.

- Check (or Study): Evaluate and monitor the results of the tests and preliminary implementation, including data measurement collection to assess whether the actual outcomes change and meet the expectations based on the KPI and objective defined in the planning step.
- Act (or Adjust): Implement solutions fully based on what was learned in the Check/Study step, identify, and sustain improvements. And then document lessons learned, update standards as needed, and set a new baseline for the next PDCA cycle.

The above description of the four steps above is a common practice [Tague 2005]. There are variations of PDCA. For instance, one practice in the "Do" stage is to fully implement solutions. Then, in the "Act" stage, the efforts focus on standardizing and stabilizing the achievement made [LEI n.d.]. Seemingly, different versions of PDCA arrange PS and CI efforts in different stages, while the overall flow is consistent. From this perspective, different versions follow the same principle and should have the same results if applied effectively.

The four steps P-D-C-A can be mapped to the three phases of PS and CI, as depicted in Figure 6.7. In a PS and CI project, the four steps are typically documented in a standardized form for sharing and updating.

PDCA has been extensively applied across various industries, including in tissue manufacturing [Chikwendu et al. 2020], manual assembly [See 2022], and furniture production [Csanády et al. 2019].

6.1.2.4 Approach Selection

Interestingly, these approaches have their own steps and features, which reflect their focuses and considerations. Table 6.1 summarizes the key features of the four approaches for reference when selecting an approach.

Moreover, their steps and efforts can be summarized into three common phases, following the same overall process flow. For comparison purposes, Figure 6.8 illustrates the four popular approaches to the three fundamental phases, which are shown at the bottom of the figure. Overall, these approaches are very similar.

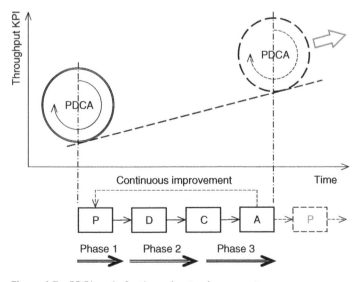

Figure 6.7 PDCA cycle for throughput enhancement.

Table 6.1 Key features of common approaches.

Approach	Key feature
DMAIC	Structured data-driven approach, emphasizing data and analysis
8D/5D	Team-focused, facilitated by a standardized report format
A3	In one-page format, concise communication and documentation
PDCA	Cyclical framework for continuous improvement

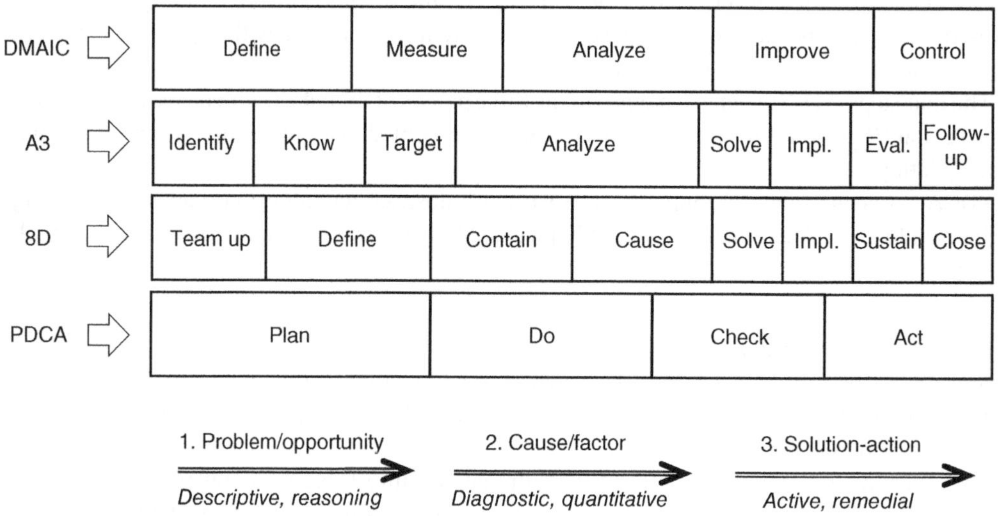

DMAIC ⇨	Define	Measure	Analyze	Improve	Control

A3 ⇨	Identify	Know	Target	Analyze	Solve	Impl.	Eval.	Follow-up

8D ⇨	Team up	Define	Contain	Cause	Solve	Impl.	Sustain	Close

PDCA ⇨	Plan	Do	Check	Act

1. Problem/opportunity ⟹	2. Cause/factor ⟹	3. Solution-action ⟹
Descriptive, reasoning	*Diagnostic, quantitative*	*Active, remedial*

Figure 6.8 Comparison of common problem-solving approaches.

Table 6.2 Methods of maintenance improvement.

Approach	Method
Principle	TPM: Total productive management
	ALM: Asset lifecycle management
	TQM: Total quality management
	Kaizen (Six Sigma): CI, variation reduction
	Lean manufacturing: Lean principles based maintenance
Analytical method	FMEA: Failure mode and effect analysis
	QFD: Quality function deployment
	FTA: Fault tree analysis
	Benchmarking: Comparison to industry best practice
	ABC analysis: Inventory significance categorization
	SWOT analysis: Strengths, weaknesses, opportunities, and threats
Simple method	Cause–effect analysis
	Pareto analysis
	5S: Sort, set in order, shine, standardize, sustain
	5W1H: Who, what, where, when, why, how

However, there is no consensus on which approach is best for every situation. A team or organization often has its own view on how to select a more "appropriate" approach. Therefore, choosing an approach depends on judgment, as all can be effective.

In different areas, professional organizations may recommend some PS and CI tools. For example, in maintenance, the Association of German Engineers's VDI 2887 suggests some methods [VDI 2019], refer to Table 6.2.

When it comes to PS and CI, one approach is normally sufficient. However, it may be beneficial to combine or adapt different approaches depending on the situation and the needs of the stakeholders. Familiarity with all common approaches can help apply one approach effectively. In addition, several other similar approaches have been developed and used in different fields and contexts.

Subsection Summary: This subsection examines four common PS and CI approaches: DMAIC, 8D/5D, A3, and PDCA, on their shared fundamental three-phase structure. As all approaches can be effective, the choice of approach relies on team preferences.

6.2 Core Process for Throughput Enhancement

6.2.1 Throughput Enhancement Essentials

6.2.1.1 Data Driven and Monitoring

Data-driven is a foundation for throughput management. A quote, often attributed to Dr. W. Edwards Deming although its exact origin is unclear, states: "Without data, you're just another person with an opinion." All steps in PS and CI, including brainstorming, analysis, reasoning, solution, and verification, require data. However, it is important to note that intuition and experience also play significant roles in data collection, analysis interpretation, reasoning, and decision making in the throughput enhancement process.

Applying a data-driven approach may present various challenges. For instance, sensors and monitors supply plentiful operational data. However, data reliability and specificity can be questionable. It is important to verify and validate automatically collected data before analyzing it to identify throughput issues, bottlenecks, and root causes. Another example is manual operations, such as job shop processes, which often lack sufficient data on throughput status and equipment reliability. The data collected can be manual recording, lack of consistency, and verification. Therefore, the first step in improving operational performance is to obtain reliable data, which may require additional tasks.

To support a data-driven approach in throughput management, production performance monitoring should include key capability functions, such as:

- Detect operational status deviations
- Monitor multiple KPIs
- Graphically display current status

Through basic analysis, a monitoring system should be able to:

- Diagnose deviations and discrepancies
- Suggest corrective action hints
- Predict future deviations or issues

Figure 6.9 shows an example of a KPI-based monitoring system. When quality information is available online, the production monitoring system can display quality status and OEE on

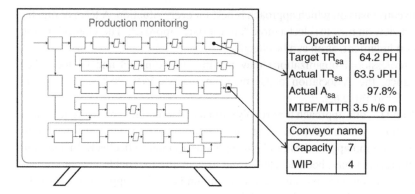

Figure 6.9 Display of production throughput monitoring.

its screen. It should display the KPI status of the system, including related conveyor systems, using color. Data display can be organized into two or three layers. Reliable and abundant data in monitoring systems can facilitate expedite and streamline throughput enhancement.

Expanding on a data-driven approach with reliable data, the next important consideration is how to select appropriate data and metrics in improvement processes.

6.2.1.2 Process Flow and KPIs

As discussed in the previous subsection 6.1.2, several structured approaches, including DMAIC, can be used for throughput enhancement. When choosing DMAIC, its five steps can be customized for specific throughput enhancement work, as depicted in Figure 6.10. With the same rationale, the 8D and A3 approaches can also be customized for throughput enhancement, adapting their steps to focus on identifying and resolving bottlenecks. While the overall process remains consistent, throughput projects place a strong emphasis on identifying and resolving bottlenecks, for instance, in the second step "Measure" in the DMAIC process and "D4" in the 8D process.

The choice of appropriate KPIs can significantly impact the accuracy of bottleneck identification. For operational availability issues, using standalone availability (A_{sa}) or standalone throughput rate

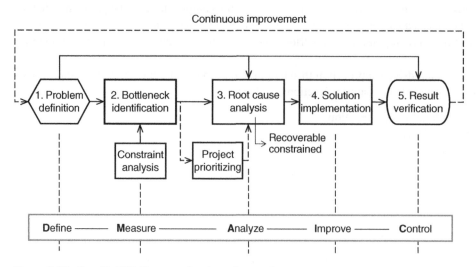

Figure 6.10 Specific DMAIC process for throughput enhancement.

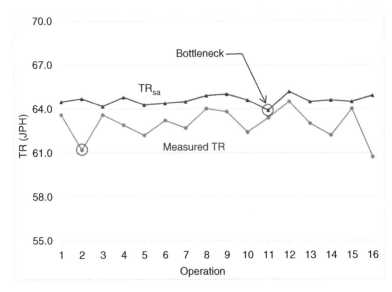

Figure 6.11 Throughput enhancement target based on TRsa.

(TR_{sa}) is recommended. Because these KPIs exclude the influences from upstream and downstream systems, providing an accurate representation of system performance, as discussed in Section 3.2 of Chapter 3. Figure 6.11 provides an example of both TR_{sa} and measured TR data of a system for discussion. Operation 11, identified with the lowest TR_{sa} value, serves as the bottleneck operation of the system. Conversely, using measured TR incorrectly points to Operation 2 as the bottleneck.

Addressing multiple KPIs is often necessary in manufacturing throughput projects. In the example discussed in Chapter 3 (Figure 3.10), four KPIs, including TR, MTBF, MTTR, and scrap ratio, are used to measure its performance. Typically, addressing multiple KPIs needs prioritization for improvement directions. The significance of each KPI and the required efforts can vary in terms of scale and investment requirements. Some KPIs can influence each other (as discussed before), further complicating prioritization.

When the focus of an improvement project is on improving product quality, quality-related KPIs should be used to identify the bottleneck and track the project's progress. In addition to quality KPIs, comprehensive metrics like OEE can also be used because OEE is partially influenced by a quality KPI.

Subsection Summary: In manufacturing throughput enhancement, a data-driven approach with reliable data is essential. The choice of KPIs, such as TR_{sa} for system performance, is crucial to correct bottleneck identification.

6.2.2 Bottlenecks and Root Causes

6.2.2.1 Bottleneck Focus

As a key distinction for throughput enhancement from the general PS and CI processes, focusing on system bottlenecks means that any addressed problem or opportunity should be related to throughput bottlenecks based on TOC (Theory of Constraints). This logic aligns with the bottleneck analysis, as thoroughly discussed in Chapter 3. Here, we provide a recap and delve further into the importance of bottleneck identification for the success of throughput projects. Without this crucial step, subsequent actions, like root cause analysis, may not yield meaningful results.

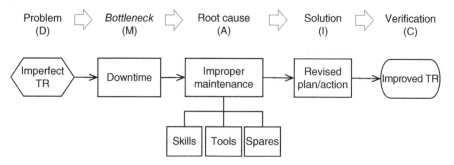

Figure 6.12 Problem-solving process for throughput improvement.

For bottleneck identification, there are several useful methods, including operational availability, cycle time, active period, standalone OEE, buffer WIP analysis, and more. The TOC provides a fundamental principle for identifying bottlenecks. Full adherence to the five TOC steps (as shown in Figure 3.2) may not always be necessary, particularly when bottlenecks are easily identifiable.

Once the bottlenecks of a system are identified, the associated losses in terms of TR, quality rate, and/or other relevant metrics can be monetized. This information on the losses and improvement investments informs management decisions to prioritize plans, as detailed later in this chapter.

Figure 6.12 illustrates a five-step downtime-focused improvement process flow, which follows the DMAIC process shown in Figure 6.10 as an application example of DMAIC.

Sometimes, root causes can be apparent. For instance, the lack of equipment maintenance is often recognized as a direct reason for downtime, serving as a bottleneck limiting system throughput. Aside from equipment repair, maintenance management may require improvements. Several key factors must be addressed, including maintenance scheduling, the availability of skilled trades, maintenance tools, as well as spare parts and materials for maintenance. Once the identified root causes related to maintenance are resolved, a revised maintenance plan can be developed to prevent the recurrence of the same and similar downtime issues.

As bottlenecks can shift their locations and magnitudes over time, as discussed in Chapter 3, it is important to continuously monitor and update bottlenecks during PS and CI projects. This underscores the sustained improvement efforts and the need for ongoing analysis and evaluation of manufacturing processes.

6.2.2.2 Drill-Down Process

Once bottlenecks are identified, the next task is to identify the corresponding root causes. In large, complex systems, root causes may not be obvious. In such situations, conducting a drill-down analysis can be an effective approach to analyzing specific issues, such as downtime, slow cycle time, etc., into the basic elements of a system and determining the root causes. The drill-down analysis process, as a general approach, is applicable in various disciplines, such as business [Rademacher et al. 2022] and data management [Rhodes 2022].

A drill-down process is exemplified in Figure 6.13. The process follows the hierarchical structure of a manufacturing system, from production line to workstation to equipment. After tracing the primary causes of the system bottleneck to specific equipment, data on the duration and frequency of downtime for that equipment can be gathered. Subsequently, equipment parameters should be carefully examined.

During a drill-down exercise, it is important to focus on the most critical issues, since there are often multiple challenges to address. Pareto analysis and charts are effective tools for visualizing

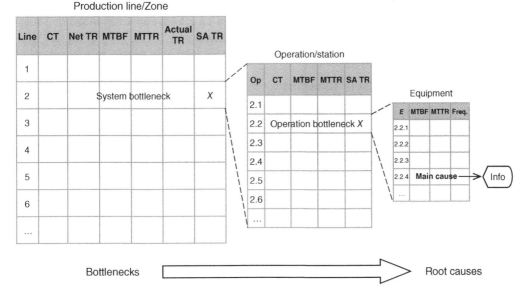

Figure 6.13   Drill-down process for root cause analysis.

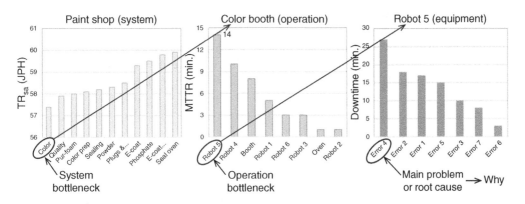

Figure 6.14   Example of a drill-down process to root causes.

the ranking of multiple bottlenecks. Figure 6.14 provides an example of a drill-down exercise in a vehicle paint shop. In this case, the system bottleneck is the Color line. Within that line, Robot 5 serves as the bottleneck, and the root cause is identified as Error type 4.

A deeper investigation into root causes may be necessary, examining conditions or factors that lead to Error 4 and understanding the underlying reasons. Such an analysis and findings are also valuable in preventing the recurrence of these issues.

In addition to addressing system downtime, drill-down analysis applies to other KPIs. For cycle time improvement, specifically, the exercise would involve progressing from the production line level down to individual workstations, machines, and process steps.

As discussed in Chapter 3, throughput issues are categorized into two fundamental types: capacity bottlenecks and performance bottlenecks. Understanding the root causes more related to bottleneck types promotes and enables targeted solutions.

Subsection Summary: Reviewing KPIs, processes, and considerations for bottleneck identification, this subsection introduces the drill-down analysis as an effective tool for uncovering root causes, for targeted solutions for system throughput excellence.

6.2.3 Variation Management in Throughput

In many situations, issues can vary in location, amplitude, characteristics, etc. Therefore, the throughput management should consider not only the average value of a KPI but also its variation over time. Dealing with variation requires specific considerations regarding its nature and impact. From this standpoint, managing variation based on a solid understanding and analysis of its characteristics is foundational for throughput enhancement.

6.2.3.1 Understanding Variation

Operational performance inherently varies. As Dr. W. Edwards Deming said, variation is life and life is variation [Deming 1994]. Even in automated manufacturing operations, KPI deviations are common, with varying impact and source. Understanding the origins of variation helps identify root causes.

To effectively address variation, three key questions must be asked. The first question concerns the measurement of variation. Three common metrics are:

- Range (R): Defined as the difference between the maximum and minimum values in a dataset. The range is preferred when dealing with data sets of fewer than five observations.
- Standard deviation (σ): The square root of variance, easily calculated using a computer program like Excel's "stdev.s" function. σ is suitable when dealing with five or more observations.
- Coefficient of variation (CV): Defined as the ratio of the standard deviation (σ) to the mean (μ) of given data.

$$CV = \frac{\text{standard deviation}}{\text{mean}} = \frac{\sigma}{\mu}$$

The second question concerns amplitude. In many cases, there is an acceptable variation level, depending on the attributes being measured and working conditions. For instance, in a production operation, CV <0.75 is considered indicative of low variation and CV >1.33 indicates high variation [Hopp and Spearman 2011]. Another example of KPI is the process capability index (C_p), calculated as the ratio of a product specification to six times the standard deviation. A C_p value of ≥1.67 is considered excellent.

The third question concerns identifying the origins and sources of variation, which can be broadly categorized into two types: common causes and special causes. Common causes are inherent in a system, sometimes referred to as "natural causes" or random variations in an operation. This type of variation is largely determined by system design. For throughput improvement on the production flow, the focus should be on special causes.

6.2.3.2 Variation from Special Causes

Special causes originate from operational situations and process parameters, such as out-of-spec environmental conditions, a lack of operator training, and inadequate equipment maintenance. Special causes frequently result in outliers and/or a trend in KPIs for operations. Also known as "assignable causes," they can be removed or corrected on the production floor.

In the context of a throughput PS and CI project, it is crucial to determine if a large variation exists in the system during bottleneck identification. Large variation might suggest the influence of several special causes during that period in question. Due to the presence of large variation, an action is less likely to produce consistent results because it may resolve one of the existing root causes. Therefore, the first focus should be on addressing special causes and reducing process variation.

Determining what constitutes "large variation" depends on the specific context, such as the type of process and quality requirements, and can be informed by historical data. Here are a few reference examples:

- Range (R) should be <10% of the mean.
- Standard deviation (σ) should be <12% of the required specifications.
- Coefficient of variation (CV) should be <1.33.

Once the known major special causes have been addressed and resolved, the associated outliers, trends, or large variation should no longer be present. Consequently, KPIs with low variation can reliably indicate bottlenecks.

When it is challenging to identify the reasons behind outliers, it is sometimes reasonable to exclude them and/or analyze KPI data over a more extended period. In these cases, with outliers removed, a KPI may better represent the status as a basis for improvement.

Subsection Summary: Managing throughput enhancement projects requires a comprehensive approach encompassing the understanding, measurement, and analysis of variation. Special cause variations should be promptly addressed and resolved. Effectively managing variation enables reliable KPIs for throughput improvement.

6.3 Particular Throughput Analysis

6.3.1 Downtime Analysis

6.3.1.1 Downtime Behavior

Downtime refers to periods when a system is nonoperational due to failures. Such failures result in lost production time and hinder system throughput. Therefore, minimizing downtime is often a top priority in throughput management.

System elements, including equipment, tooling units, and processes, are designed to operate reliably for extended periods. However, even with proper maintenance, these can still experience random failures. Most failures are short-lived, typically lasting less than one or two minutes for mass production, but can occur frequently.

Observations suggest that system downtime often follows a negative exponential distribution. Downtime may follow other distributions, such as Weibull and lognormal. Curve fitting, as demonstrated in Figure 6.15, mathematically describes downtime patterns. The resulting mathematical expression provides valuable information for maintenance planning.

In real-world cases, curve fitting may not yield a perfect fit and involves some approximation. Curve fitting accuracy can be quantitatively measured. For instance, when using MS Excel, the accuracy of a fit is assessed using R^2, known as the coefficient of determination. A higher R^2 value indicates a better fit. An R^2 value of 1 signifies a perfect fit; in practice, however, values above 0.8 are generally considered acceptable. This implies that the fitted function can explain approximately 80% of the variation in y based on x. Figure 6.16 provides an example of curve fitting applied to downtime data.

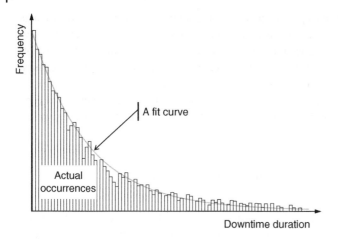

Figure 6.15 Curve fitting of system downtime distribution.

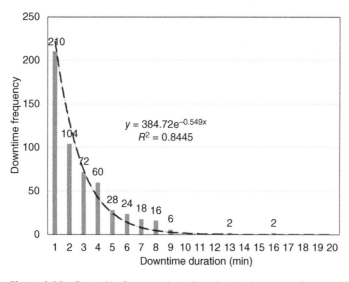

Figure 6.16 Example of system downtime data and exponential curve fitting.

6.3.1.2 Major and Minor Downtime

The duration and frequency of downtime vary depending on several factors, including system and equipment characteristics. Figure 6.17 illustrates three scenarios with different downtime durations, assuming exponential distribution function curves.

Comparing the three downtime curves in the figure, the main difference is the spread of most downtime events, showing 99% in a specific period, in this example. Thus, downtime can be categorized as major and minor. Their definitions depend on the context, situations, and expectations. For instance, for high-volume production, major downtime events last over 15 minutes, and minor downtime lasts five minutes. The downtime between can be treated based on other factors, such as the impact on neighboring operations.

Major downtime is rare but lasts for an extended period, surpassing buffer capacity. Consequently, addressing major downtime should be a top priority and involve measures aimed at preventing its recurrence.

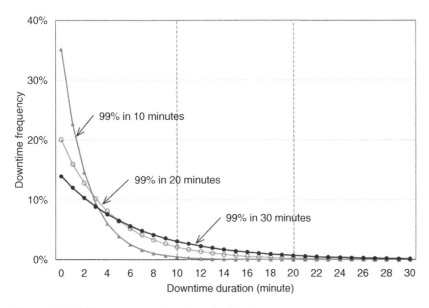

Figure 6.17 Different system downtime distributions.

Minor downtime, as discussed before, may be absorbed by system buffers, resulting in limited or minor impact on system throughput. However, if a system has frequently repeating minor downtime, its cumulative time lost due to frequent minor downtime can be significant, exceeding buffer capability. Accordingly, such minor downtime should be addressed.

Furthermore, downtime patterns tend to stay consistent under stable production operations. In such scenarios, historical data can be used to predict future downtime situations under the same conditions. Assuming stable conditions, system throughput and downtime status can be analyzed and predicted using various mathematical models, such as the moving average (MA) model, auto-regressive (AR) model, and their combination (ARMA). Several studies have used these models, including Johnson et al. [2022] Subramaniyan et al. [2019] Huang et al. [2017], and Li et al. [2011]. These studies provide insights into downtime behavior and benefit maintenance management.

Subsection Summary: Downtime can be analyzed using curve fitting to characterize distributions. Major and minor downtime events, in terms of time, have different impacts and require different considerations. Analyzing downtime provides valuable insights for throughput and maintenance management.

6.3.2 Cycle Time Analysis

6.3.2.1 Cycle Time Measurement

As defined and discussed in subsection 1.2.2 in Chapter 1, CT reflects the operational pace of an operation. Every manufacturing operation has a design (required) CT. A CT issue refers to when the actual work is slower than what was required. Even though the delay in each cycle may be small, its impact can be significant. This is because CT discrepancies occur in every cycle and accumulate over an extended period.

While a subtle challenge with CT issues is that they, unlike downtime issues, may not be noticeable without careful measurements because they do not involve a complete stoppage. Therefore,

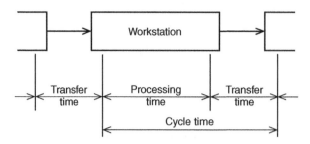

Figure 6.18 Components of workstation cycle time.

CT issues represent a hidden threat to system throughput. Finding and addressing CT issues begins with CT measurement.

Here are some points regarding CT measurement, in addition to the discussion in Chapter 1 and shown in Figure 1.9, to ensure CT measurement accuracy measure:

- Ensure to take measurements at consistent points, such as unit entry or exit, for each cycle.
- Measure multiple times and average over 5–10 consecutive cycles.
- Exclude any stoppages caused by blockages and starvation.
- Disregard abnormal (larger than normal) gaps between the units, regardless of the reasons.
- Include internal transfer time for each operation, if applicable, as illustrated in Figure 6.18.

Once obtaining CT measurements, one can compare their mean to the design CT and identify whether the measured CT is slower and requires improvement. To identify CT issues in a system, measuring and analyzing cycle time of all workstations, at least target workstations, across an entire system.

6.3.2.2 System Cycle Time Analysis

As discussed in Chapter 1, the system CT is determined by the CT of the slowest operation/workstation when a system is in serial configuration (refer to Figure 6.19). This calculation assumes that all operations are perfectly reliable and fully independent, as explained in Chapter 8. When the assumptions do not hold true, the result is lower than the theoretical value.

$$CT_{serial} \approx Max\{CT_{op.1}, CT_{op.2}, \dots CT_{op.n}\}$$

For parallel systems (Figure 6.20), the system's CT is about the algorithm average of the CT of the parallel segments, divided by the number of segments, as shown in the equation below. Note that the parallel segments are assumed to have similar capacity and operate with perfect reliability. If the parallel segments have different production volumes, it is necessary to consider corresponding weights based on the volume of each segment.

$$CT_{parallel} \approx \frac{Average\{CT_{seg.1}, CT_{seg.2}, \dots CT_{seg.n}\}}{n}$$

A complex system can consist of multiple serial and parallel segments. To apply CT analysis approaches to serial and parallel systems, a complex system needs to break down into a series

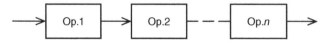

Figure 6.19 System with multiple operations in a serial configuration.

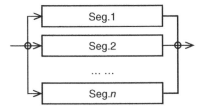

Figure 6.20 System with multiple segments in a parallel configuration.

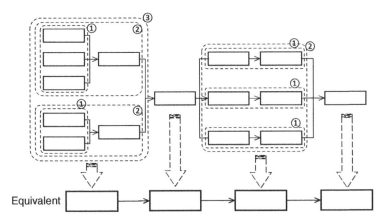

Figure 6.21 System segment consolidation for cycle time analysis.

of serial segments, as illustrated in Figure 6.21. The equivalent model can be obtained through a three-step process:

1. Consolidate small serial and parallel segments, marked as ① in the figure.
2. Consolidate these small segments with their neighboring segments, marked as ② in the figure.
3. Consolidate the segments from the previous step, marked as ③ in the figure, and repeat, as necessary.

The resulting equivalent system is serial and presented at the bottom of the figure. Then, the CT analysis for a serial system can be applied.

Subsection Summary: This subsection examines the CT analysis in throughput management, focusing on CT measurement considerations and analyses in serial, parallel, and complex systems. These insights contribute to a solid understanding of CT concepts and applications in throughput management.

6.3.3 Human Factor Consideration

6.3.3.1 Human Factor Analysis

Human factors, encompassing aspects like safety, ergonomics, efficiency, and training, play a pivotal role in manufacturing performance. Key interconnected approaches to the study of human factors in manufacturing include:

- Human working environment: Focuses on designing and maintaining a safe and ergonomic-friendly environment for human workers to enhance their productivity. This involves designing workspaces, tools, and procedures with a focus on ensuring worker safety and comfort.

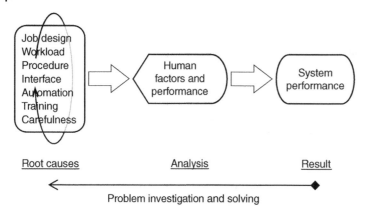

Figure 6.22 Human factors in system throughput performance.

- Operator issues and error management: Addresses potential operator issues and errors during manufacturing operations. This involves not only addressing errors when they occur but also implementing preventive measures to reduce the likelihood of errors.
- Involvement of production workers: Involves actively engaging production workers and skilled tradespeople in efforts to improve system performance. For instance, this can be achieved through practices like autonomous maintenance (AM) within TPM discussed in Chapter 5. This approach encourages workers to take ownership of equipment and operations.

Research demonstrates that a comprehensive understanding and management of human factors are crucial for optimizing manufacturing processes and ensuring the well-being of the workforce. Substantial research exists on the application of human factors within manufacturing, highlighting the significance of these considerations in industry practice.

An effective approach is looking for system and design root causes, rather than blaming solely particular issues and operators, when some human factors impact productivity or quality. Operator errors or issues are often a result of a complex interplay of multiple factors. These factors encompass various aspects, including training, process design, operational procedures, automation interface, etc. (refer to Figure 6.22).

These factors can be the real root causes of human performance issues, and addressing them is necessary to prevent the same issues from happening again. However, it is important to recognize that human performance issues are often intricate and may not have straightforward or absolute solutions. Further discussion will expand on this in the next subsection 6.3.3.2.

6.3.3.2 Challenges with Human Factors
Analyzing human factors in operational management to enhance throughput performance can be a complex task and presents several key challenges that need careful consideration. These considerations include:

- Individual variation: Operator performance can vary significantly due to differences in their capabilities, experiences, skills, and other personal factors. Even with well-defined procedures, individuals may execute tasks differently. Operational management must acknowledge and account for these variations.
- Interactions between human factors: Human factors often interact with one another, leading to complex situations. Performance issues can result from the simultaneous interaction of multiple variables. For example, human errors may be influenced by factors such as inadequate training,

a high workload, and reduced attention, which are all interconnected and demand a holistic approach.

- Influence of external factors: External factors can affect human performance. For instance, the absence of clear and measurable criteria for evaluating manual operational performance, flawed procedures, or poorly defined instructions may cause confusion and contribute to human errors. All these can ultimately affect overall throughput.
- Resistance to individual performance evaluation: Focusing on individual performance may sometimes face resistance, especially in situations where workers have union representation or in areas with legal protections. Examining manual operations may require gaining consent and cooperation.

Furthermore, manual operation speed, or the speed performance (P) element in OEE, is influenced by several factors. Time studies are valuable tools for delving into manual operations and uncovering opportunities for improvement. A comparable analysis should also be applied to safety and ergonomic issues, ensuring an assessment of the operational processes. When conducting a time study, it is important to adhere to Methods-Time Measurement (MTM) guidelines, which have evolved over decades and exhibited variations across regions and industries [Nicolle et al. 2022]. Based on a time study, CI initiatives can focus on identifying root causes.

In light of these challenges, enhancing throughput improvements necessitates an understanding of the intricate human factors involved in any throughput analysis. Managing these challenges and factors requires both an art and a science. With a solid understanding of these factors, operational management can craft and implement targeted strategies to optimize both individual and team performance.

Subsection Summary: This subsection underlines the analysis approaches of human factors and considerations. It acknowledges the challenges associated with human factors, like variation, factor interactions, and external influences.

6.4 Project Proposal Management

6.4.1 Proposal Development

6.4.1.1 Elements of Improvement Proposals

In a manufacturing environment, throughput PS and CI activities should be considered as routine work or a habit. Accordingly, throughput PS and CI activities can be categorized into two groups: simple and major. The simple ones are routine, relatively small in scope, taking a short time to complete, and without involving extensive collaboration with other departments. Such simple PS and CI activities may not require formal documentation for preparation, planning, approval, and execution. However, after completion, summarizing the actions taken and the resulting differences is necessary.

In contrast, major or larger-scale PS and CI activities are considered formal projects as they require certain resources and extended periods, for example, weeks or even longer, to complete. Such PS and CI proposals should be well prepared and formally documented. Operations management must review these proposals for evaluation, prioritization, and decision making.

Major throughput PS and CI proposals can be prepared in a standardized format, making them more accessible and streamlined for review and comparison, rather than a lengthy report. Figure 6.23 provides an example of a standardized throughput improvement project proposal form. The five elements listed in the form are essential for project evaluation.

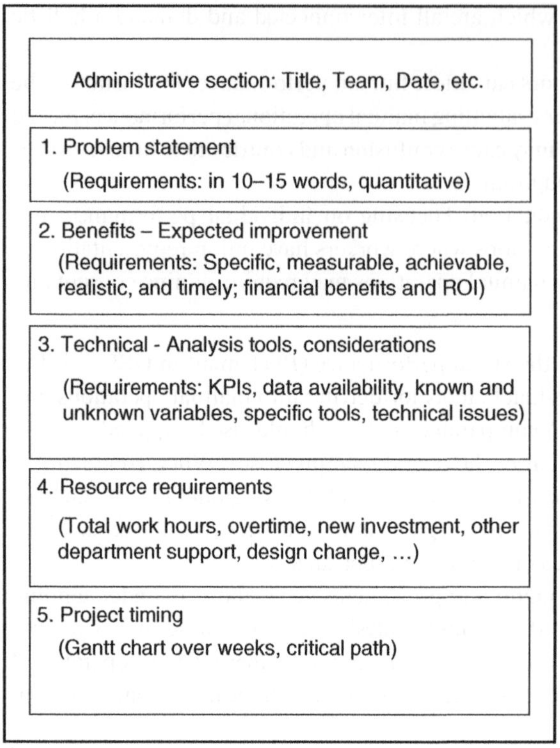

Figure 6.23 Example of throughput project proposal form.

A main challenge for a throughput project is often the allocation of resources, particularly capital investment. As introduced in Chapter 2 and discussed in several chapters, throughput accounting can demonstrate the financial benefits, including created values, documented savings, and improved productivity, showing the financial benefits outweigh the necessary investment, refer to Section 4.2 for project economic analysis.

6.4.1.2 Application of Value Proposition

The value proposition for an improvement project proposal is crucial in defining its purpose and potential benefits. There are several approaches to crafting a value proposition, and each approach articulates how the project will create value for the target system.

One effective approach is in the "we help X achieve Y by doing Z" formula. This X–Y–Z formula explicitly defines the system (X), the desired improved performance (Y), and the specific actions (Z) to take throughout the project. This formula succinctly conveys the project's goals and anticipated results. Here is an illustrative example for reference. A project proposal could state, "the project helps the manufacturing shop improve its OEE by 0.2 percentage points through the implementation of a new preventive maintenance plan within two months."

Complementary to that approach, another useful tool for crafting a value proposition is the value proposition canvas, a visual framework used in business analysis [Siltala et al. 2018; da Costa et al. 2020]. This canvas considers the customer (system) jobs, pains, and gains, and aligns them with the pain relievers and gain creators provided by the project. Refer to Figure 6.24 for an example applicable to throughput improvement projects. By completing the canvas, project teams can achieve

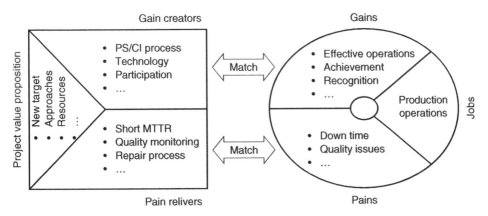

Figure 6.24 Example of value proposition canvas for improvement projects.

and present a comprehensive understanding of how their proposed project meets specific needs, addresses pain points, and generates tangible benefits to accomplish their objectives.

Using the $X—Y–Z$ formula or the value proposition canvas thinking approach can guide preparing a throughput enhancement project. By clearly defining the value proposition, the project team can better communicate its purposes and expected outcomes, ensuring alignment with the needs and goals of the system or customer. Once the value proposition is established, the next crucial step is to conduct a detailed, quantitative analysis of the specific benefits that the proposed project offers.

6.4.1.3 Considerations in Benefit Analysis

When developing the value proposition and quantifying the financial benefits through cost-benefit analysis, nonmonetary benefits such as reduced downtime and improved quality, should be translated into financial values. Methods like throughput accounting and quality cost analysis, as discussed in early chapters, can be used to quantify these economic benefits for throughput and other types of improvement.

Applying throughput accounting and/or quantitatively evaluating financial performance can face difficulties, for instance, due to data availability. PS and CI project teams should make every effort to consider quantitative financial values for their projects.

Certain benefits from PS and CI projects may be hard to quantify, especially when they involve qualitative or intangible benefits or lack a direct monetary impact (refer to Figure 6.25). For

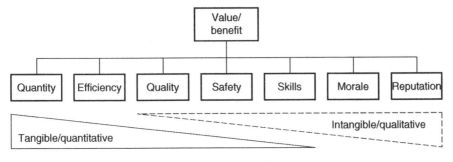

Figure 6.25 Various types of benefits from throughput enhancement.

instance, the improvement of a specific quality attribute may be hard to assign a monetary value to. Particularly for CI projects, their benefits can be considered as cost avoidance from risk reduction of certain downtime issues. In these cases, the benefits may stay descriptive rather than numerical, or they can be rated on their own merits.

In addition, an approach is to assess intangible benefits using a one to five rating scale, as discussed later in the evaluation process (Figure 6.29).

It is important to note that benefits from PS and CI projects do not carry equal importance. For instance, safety concerns or benefits typically take top priority, even though there is no clear monetary value. Another example is that quality may outweigh process efficiency in terms of significance in certain cases.

Based on this discussion, throughput enhancement has multiple dimensions of values and benefits that go beyond mere quantity or output rate. Comprehensively considering these values and addressing intangible benefits are crucial when selecting throughput enhancement projects.

Subsection Summary: This subsection delves into the creation of throughput improvement proposals, discussing their fundamental elements, introducing approaches for value proposition, and exploring the methods for addressing intangible benefits in PS and CI projects.

6.4.2 Simple Proposal Approaches

The need for system throughput improvement is often self-evident. Deciding the most pertinent targets for throughput improvement can be challenging, especially in situations with limited data or widespread performance issues. However, two straightforward approaches can be useful: a quick survey and issue monitoring.

6.4.2.1 Quick Survey

In some cases, there may be no clear direction based on KPIs to pursue to enhance system throughput. In such a situation, conducting a survey of relevant professionals can yield valuable insights. The aim of this survey is to collect diverse opinions and identify gaps and priorities in improvement directions.

The survey can be completed in four steps, illustrated in Figure 6.26, and can be completed within a few hours.

The first step is to create a few questions, addressing specific aspects of the throughput status and concerns, such as:

- Is the current maintenance sufficient to reduce MTTR?
- Can the quality defect rate be reduced by improving incoming parts?
- Are the known bottlenecks addressed?
- Is it necessary to invest in monitoring and sensing of operations?
- Is more worker training needed?

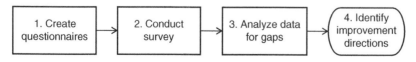

Figure 6.26 Survey process of gap analysis for improvement direction.

Table 6.3 Case of gap analysis for maintenance improvement (Maletič et al. 2014/with permission of Emerald Publishing Limited).

Aspect	Importance	Agreement	*Gap*
1. To keep a low inventory of spare parts	3.6	3.0	0.6
2. To reduce repair time	3.6	3.2	0.4
3. To help improve the production process	3.6	3.0	0.6
4. To perform periodic replacement	2.6	2.6	0.0
5. To record quality rate	3.8	3.6	0.2
6. To improve the skills of maintenance staff	3.8	2.4	1.4
7. To be based on statistical model and data	2.6	1.2	1.4
8. To analyze equipment failure causes	3.6	2.8	0.8
9. To use computer maintenance systems	3.2	2.0	1.2
10. To monitor equipment status	4.2	2.4	1.8

Each question is asked twice, using a five-point Likert scale:

1. On the agreement of this aspect: (1) Strongly disagree, (2) Disagree, (3) Neither agree nor disagree, (4) Agree, and (5) Strongly agree.
2. On the importance of this aspect: (1) Not important, (2) Slightly important, (3) Important, (4) Fairly important, and (5) Very important.

After receiving the survey results, the third step is to analyze them. For each question, there are two average scores. The difference between the two average scores is considered as the gap for the aspect, where the larger gaps signify the respondent's opinions about which areas to improve first. A case study in a textile company was conducted, as shown in Table 6.3. For instance, #10 is the top priority from the table.

This approach is good as it collects opinions. Because of the subjective nature of the gaps based on observations and experiences, further analysis along with other types of data, such as KPI monitoring and financial information, can result in more reliable directions to go.

6.4.2.2 Standardized Issue Tracking

It is common that throughput issues are documented by different teams, totaling over a hundred. These documented issues likely point in different directions for throughput improvement. Organizing and consolidating these issues can create comprehensive views for effective communication, identifying concerns to prioritize, encouraging deeper analysis, and follow-up. This practice is common and effective on the production floor.

A standardized form or spreadsheet, like Table 6.4, captures all types of throughput issues, their related information, and corresponding responses. During or after PS, discussions can explore potential CI opportunities listed in the final column, transitioning from addressing immediate issues to planning future projects. Note that the form may need customization based on the specific needs and context of the operations.

For instance, when responding to a failed sensor, the initial action is typically sensor replacement. However, further considerations might include upgrading to a more reliable sensor, relocating its

Table 6.4 Throughput issue register and tracking form.

No.	Issue	Type	Location	Severity	Owner	Target date	Updated status/date	Document (link)	Further CI opportunity
1									
2									
...									

position, or exploring alternative sensing scenarios. Another example pertains to quality-related throughput issues, where if the root cause is linked to process capability, a potential project may involve making process changes.

An advantage of using spreadsheets for throughput issue registers is that the issues can be sorted and filtered based on different criteria. That helps prioritization and follow-up for throughput management. Minor or readily fixed issues with little likelihood of recurrence do not require in-depth projects. However, critical and persistent throughput issues requiring substantial effort to resolve are categorized as major items and candidates for formal improvement projects.

As discussed earlier, throughput improvement should be data-driven and focused on identifying bottlenecks. These approaches, including the issue register form, support a systemic approach, especially in local, process-specific, and CI activities.

Subsection Summary: This subsection introduces two accessible approaches in throughput improvement projects: a quick survey of professionals and coordinating throughput issues. These methods are straightforward and can facilitate coordination in throughput improvement.

6.4.3 Proposal Review

6.4.3.1 Limit Assessment
In Chapter 1, subsection 1.3.3 discusses three major factors influencing throughput performance: system design, production planning and control, and CI. While production floor personnel often collaborate to enhance system throughput through integrated efforts in production planning and CI, the impact on issues primarily rooted in system design is limited.

To ensure the feasibility and success of throughput enhancement projects, it becomes crucial to assess the limits to improvement in terms of feasibility and effectiveness. These limits shape the path to throughput excellence. Figure 6.27 illustrates an example, showcasing the improvement potentials of the individual factors toward ideal throughput excellence.

Figure 6.27 illustrates a general principle for large, complex systems. In smaller systems or specific operations, only one or two factors, such as processes and machines, may be relevant for improvement.

Limits can be identified during the root cause analysis of bottlenecks, as depicted in phase 2 (root cause analysis) of Figure 6.8. Root causes are then considered into two parts: those resolvable on the floor and those fixable by design changes. For example, while on-floor measures can improve slow CT by 5–10%, larger improvement likely needs design changes.

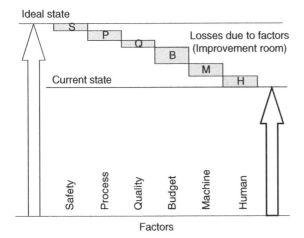

Figure 6.27 Limited room for throughput enhancement.

6.4.3.2 More Considerations

Beyond limits and root causes, additional considerations enhance throughput improvement projects in complex systems:

- Impacts and prioritization: A Pareto chart can illustrate distinct levels of impact that root causes have on throughput enhancement. While prioritizing the most impactful root causes is straightforward, feasibility could be constrained by technical and financial conditions. In such cases, a practical strategy is to tackle the most resolvable root cause, or a "low-hanging fruit," for a quick improvement and prepare to address the most impactful later.
- Connection and interactions: Root cause analysis should consider interactions among system elements, KPIs, and related root causes. They can often interact in complex ways, such as throughput rate, maintenance, and resources. Balancing and prioritizing efforts and establishing realistic goals become feasible by reviewing the individual significance and relationships among these elements. Low-feasibility solutions can be postponed until circumstances become more favorable.
- Quantification of root causes and targets: Quantifying constraints with data on resource limitations, budget impacts, etc. For instance, a resource constraint might translate to a 0.5 JPH improvement with a $100,000 investment in maintenance functions. This quantification is crucial to demonstrate to management to approve measures overcoming budget constraints and determine realistic expectations for improvement.

6.4.3.3 Project Proposal Evaluation

Evaluating throughput enhancement proposals can be a complicated process due to numerous factors and considerations. The throughput management team assesses and prioritizes all proposed projects based on three key aspects: (1) benefits, (2) technology, and (3) resources.

1. The benefits of addressing throughput losses can be quantified in monetary terms, as previously discussed. Benefits sometimes reflect other concerns, such as safety and environmental issues, which may not be easily measured in monetary terms. Furthermore, a benefit evaluation must address the benefits of the entire system, even though corresponding solutions may apply locally.

2. The technological aspect is technical readiness, which can be uncertain, particularly for unproven yet new technologies. Fully evaluating technical risks is imperative. The technical aspect is also related to financial investment and should be evaluated with financial justification.
3. The resources involve capital investments and the workforce, which influences project feasibility and timeline. Resources can be easily presented or converted into monetary terms.

For assessing the potential of throughput enhancement proposals, management may rate the proposed projects on a scale of five, where five indicates the most favorable choice. Table 6.5 illustrates an example of four proposals evaluated. The table's overall rating is derived from three ratings, exemplified by $4 \times 4 \times 2 = 32$ for proposal A.

These overall ratings enable a direct comparison of the projects. In this example, the overall rating of proposal B is the highest, making it the preferred option. Importantly, the accompanying financial information, which includes initial investment and operating expenditures, should be jointly evaluated. This comprehensive and straightforward comparison streamlines project evaluation.

6.4.3.4 Visualization Aids for Comparison

Visualizing project proposals using comparison charts simplifies the review and selection process. Two charts, based on the ratings of three aspects (benefits, technology, and resources), can help illustrate the relative positions among projects. Figure 6.28 illustrates an example, according to the evaluation ratings in Table 6.5. In the figure, the low scores (e.g., rating ≤ 2.5) indicating lower practicality can be shaded for visual emphasis.

If a proposal evaluation requires more in-depth analysis, additional aspects and detailed items can be examined. These may include sponsorship, benefits, resources, efforts, deliverables, time, team, project charter, and the proposed approach, each receiving individual scores and weights [Pyzdek and Keller 2018]. Subsequently, proposals can be ranked based on their total weighted scores.

Unlike relying solely on the product of the three aspects, this chart illustration offers a more detailed comparison. Furthermore, these three aspects are not always equally important in the context of operational excellence, immediate needs, and feasibility. Accordingly, weights may be introduced to the three aspects. The weighting approach in OEE introduced in subsection 2.3.1 can be a good reference.

Subsection Summary: This subsection discusses the evaluation of throughput project proposals, emphasizing the importance of identifying constraints, considering multiple aspects, and visualizing project comparisons for effective decision making.

Table 6.5 Example of project proposal comparison.

Proposal	(1) Benefits	(2) Technology	(3) Resources	Overall
A	4	4	2	32
B	3.5	3	4	42
C	3	3.5	2.5	26.25
D	2	4	3	24

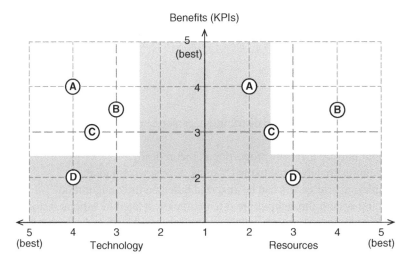

Figure 6.28 Visual comparison of project proposals across three key aspects.

Chapter Summary

6.1. Approaches for Solving and Improving

1. Problem solving (PS) and continuous improvement (CI) are structured processes applicable across fields, involving three phases: problem identification, root cause analysis, and solution implementation.
2. CI is proactive, with identified targets aiming for specific, measurable, achievable, relevant, and time-bound goals, through self-assessment or benchmarking.
3. Both PS and CI can be inductive, deductive, or a combination of both, with frequent use of the deductive process – applying general principles to specific situations.
4. DMAIC, 8D, A3, and PDCA are widely used approaches, each emphasizing unique steps within a common three-phase structure.
5. Selection depends on factors like project size, with key features including structured data-driven emphasis, team-focused format, and effective communication.

6.2. Core Process for Throughput Enhancement

6. PS and CI projects for throughput enhancement emphasize structured processes and data-driven approaches with a bottleneck focus.
7. A data-driven approach is essential, relying on reliable data, KPIs, and real-time troubleshooting for systematic improvement.
8. Processes like DMAIC should be customized for throughput enhancement, emphasizing the importance of choosing appropriate KPIs for accurate bottleneck identification.
9. Identification of bottlenecks is pivotal, and drill-down analysis helps trace root causes, essential for targeted solutions and improved system performance.

10. Variation understanding, measurement, and management are foundational, distinguishing between common and special causes to achieve accurate KPI measurement and reliable bottleneck identification.

6.3. Particular Throughput Analysis

11. Downtime events, often following exponential distribution, can be categorized into minor and major based on their frequency and duration.
12. Cycle time and throughput rate are inversely proportional, theoretically.
13. In a serial system, CT is approximately equal to the CT of its slowest workstation.
14. The cycle time of a complex, hybrid system can be analyzed based on its configuration.
15. Human factors influencing system throughput are often complex, related to various factors, such as manual operation procedures, training, machine interfaces, workload, and ergonomics.

6.4. Project Proposal Management

16. Throughput projects should include three major dimensions: value/benefits, technical readiness, and resource availability, with equal or unequal weighting.
17. Two simple approaches for proposing throughput improvement projects are a quick survey of relevant professionals and monitoring throughput issues.
18. It is important to recognize and assess the limits of KPI improvement.
19. Categorizing root causes into resolvable on the floor, constrained by design, or their combination is an effective way to solve throughput problems.
20. It is a good practice to visualize project comparisons using charts to illustrate the relative positions based on the ratings of benefits, technology, and resources.

Exercise and Review

Exercises

6.1 The TR data for a month exhibits a standard deviation of 2.7 and a mean of 56. Calculate the coefficient of variation (CV) for this data.

6.2 The table (Table 6.6) lists the downtime durations and corresponding frequencies of a system for one week. Utilize MS Excel or similar software to perform data curve fitting and analyze the overall characteristics of the system.

6.3 A serial production line consists of five workstations, each with cycle times of 55, 58, 54, 57, and 52 seconds, respectively. Estimate the cycle time for this line.

Table 6.6 For Exercise 6.2.

Downtime (<min)	1	2	3	4	5	10	15	20
Frequency (times)	35	20	7	4	2	1	1	1

Table 6.7 For Exercise 6.5.

Proposal	(1) Benefits	(2) Technology	(3) Resources
A	3	4	2
B	3.5	3.5	3
C	4	2.5	3.5

Table 6.8 For Exercise 6.6.

Task	Importance	Agreement
1	3.3	2.5
2	3.2	3.1
3	3.2	3.0
4	3.3	2.7
5	4.0	3.6
6	3.7	2.8

6.4 A parallel manufacturing system comprises three identical legs, each with an equal production volume. The cycle times for these three legs are 85, 88, and 87 seconds, respectively. Determine the cycle time for this system.

6.5 Three improvement project proposals have been rated (without weights) as shown in Table 6.7. Based on the three factors, determine which proposal should be prioritized for support. Provide a rationale for your choice.

6.6 An internal survey was conducted to assess the levels of importance and agreement for various improvement tasks, as listed in Table 6.8. Based on the survey results, identify the top three tasks that should be prioritized.

Review Questions

(The chapter covers these topics. For further discussion, it is recommended to seek additional information and examples. Diverse perspectives are encouraged.)

6.1 Compare and contrast the applications of problem-solving (PS) and continuous improvement (CI), highlighting the key differences between the two.

6.2 Do you agree that a mindset is a key factor in creating CI projects? Why or why not?

6.3 Provide an example of using either induction reasoning, deduction reasoning, or a combination of the two in a PS or CI process.

6.4 Using two common approaches, such as the 8D and A3, outline the three phases involved in a PS or CI process.

6.5 Describe the DMAIC process and provide an example of its application in a PS or CI process.

6.6 Discuss the similarities between the PDCA process and another approach, such as the 8D or A3, for an application.

6.7 Review Figure 6.8 and provide comments on the comparison of the common approaches.

6.8 In applying a PS or CI approach to throughput enhancement, what are the unique considerations in the application process? Provide an example.

6.9 Bottleneck identification should be addressed in the measurement stage of DMAIC (refer to Figure 6.10). How can we incorporate bottleneck identification in similar approaches, such as the 8D and A3?

6.10 Describe a process for drilling down root causes from the system level to the equipment level, providing an example.

6.11 Compare different measurements of variation. Explain which measurement you would recommend and why.

6.12 Explain why addressing large variations by finding their special causes should be prioritized in throughput enhancement. Justify.

6.13 Select and justify a probability distribution for modeling system downtime behavior, providing a specific example. Does the downtime follow an exponential distribution?

6.14 Provide an example of a cycle time measurement process on the production floor.

6.15 Provide an example of a hybrid system configuration and discuss how to analyze its cycle time.

6.16 Review human factors and their root causes in a production environment and discuss a CI project that addresses them.

6.17 Use an example of a PS or CI project to show its value proposition.

6.18 Do you agree that a survey of manufacturing professionals can help identify performance bottlenecks?

6.19 How can we analyze the limits and root causes of a proposed PS or CI project?

6.20 Review the three major factors (benefits, technology, and resources) involved in a throughput enhancement project, providing an example.

References

Chikwendu, O.C., Ifeyinwa, O.F., Igbokwe, N. and Utochukwu, O.E. 2020. The implementation of Kaizen manufacturing technique: A case of a tissue manufacturing company. International Journal of Engineering Science and Computing, 10(5), pp. 25938–25949.

Csanády, E., Kovács, Z., Magoss, E., Ratnasingam, J., Csanády, E., Kovács, Z., Magoss, E. and Ratnasingam, J. 2019. Furniture production processes: Theory to practice. Optimum Design and Manufacture of Wood Products, pp. 367–421.

Fernandes S. da Costa, Pigosso, D.C., McAloone, T.C. and Rozenfeld, H. 2020. Towards product-service system oriented to circular economy: A systematic review of value proposition design approaches. Journal of Cleaner Production, 257, p. 120507.

Daniyan, I., Adeodu, A., Mpofu, K., Maladhzi, R. and Kanakana-Katumba, G.M. 2023. Improvement of production process variations of bolster spring of a bogie train manufacturing industry: A six-sigma approach. Cogent Engineering, 10(1), p. 2154004.

Deming, W.E. 1994. The New Economics: For Industry, Government, and Education, 2, The MIT Press, Cambridge, MA, USA.

Ekmekci, I. and Nebati, E.E. 2019. Triz methodology and applications. Procedia Computer Science, 158, pp. 303–315.

George, A., Ranjha, S. and Kulkarni, A. 2021. Enhanced problem solving through redefined 8D step completion criteria. Quality Engineering, 33(4), pp. 695–711.

Hakala, V.M. 2018. Improvement of Production Efficiency by Using Lean Methods. https://www.theseus.fi/bitstream/handle/10024/142998/Hakala_Veli-Matti.pdf?sequence=1. Accessed April 2023.

Hardy, D.L., Kundu, S. and Latif, M. 2021. Productivity and process performance in a manual trimming cell exploiting Lean Six Sigma (LSS) DMAIC–a case study in laminated panel production. International Journal of Quality & Reliability Management, 38(9), pp. 1861–1879.

Hopp, W.J. and Spearman, M.L. 2011. Factory Physics: Foundations of Manufacturing Management, 3 ed., ISBN: 978-1577667391, Waveland Press, Long Grove, Illinois.

Huang, J.T., Meng, Y. and Yang, Y. 2017. Forecasting of throughput performance using an ARMA model with improved differential evolution algorithm. In 2017 13th IEEE Conference on Automation Science and Engineering (CASE), IEEE. pp. 376–381.

Johnson, B.J., Sen, M., Hanson, J., García-Muñoz, S., and Sahinidis, N.V. 2022. Stochastic analysis and modeling of pharmaceutical screw feeder mass flow rates. International Journal of Pharmaceutics, 621, 121776. https://doi.org/10.1016/j.ijpharm.2022.121776

LEI. n.d. Plan, Do, Check, Act (PDCA), Lean Enterprise Institute. https://www.lean.org/lexicon-terms/pdca/. Accessed August 2022.

Li L, Chang Q, Xiao G, Ambani S. 2011. Throughput bottleneck prediction of manufacturing systems using time series analysis. Journal of Manufacturing Science and Engineering, 133(2), pp. 1–8. https://doi.org/10.1115/1.4003786

Maletič, D., Maletič, M., Al-Najjar, B., and Gomišček, B. 2014. The role of maintenance in improving company's competitiveness and profitability: A case study in a textile company. Journal of Manufacturing Technology Management, 25, pp. 441–456.

Martin, G. n.d. The Meaning and Origin of the Expression: If it ain't Broke, Don't Fix it, The Phrase Finder. https://www.phrases.org.uk/meanings/if-it-aint-broke-dont-fix-it.html. Accessed August 2020.

Nicolle, C.S., Vivan, M.F. and De Oliveira, A.R. 2022. Time and motion study in the industry 4.0 era: A systematic review of the literature. Gepros: Gestão da Produção, Operações e Sistemas, 17(3), p. 1.

Pólya, G. 1945. How to Solve It, ISBN 0-691-08097-6, Princeton University Press.

Pyzdek, T. and Keller, P. 2018. The Six Sigma Handbook, 5 ed., ISBN 978-1260121827, McGraw-Hill Education.

Rademacher, D., Valdez, J., Memeti, E., Samant, K., Santra, A. and Chakravarthy, S. 2022, Modviz: A modular and extensible architecture for drill-down and visualization of complex data. In Digital Business and Intelligent Systems: 15th International Baltic Conference, Baltic DB&IS 2022, Riga, Latvia, July 4–6, 2022, Proceedings, Springer International Publishing, Cham. pp. 232–250.

Rhodes, J.M. 2022. Visualizing learning management data from SQL server using power BI. In Creating Business Applications with Microsoft 365: Techniques in Power Apps, Power BI, SharePoint, and Power Automate, Apress, Berkeley, CA. pp. 183–198.

Santos Filho, G.M. and Simão, L.E., 2022. A3 methodology: Going beyond process improvement. Revista de Gestão (Management Magazine), 30(2), pp. 147–161.

See, S.X. 2022. Improving Throughput of a Semi-automated Assembly Line Subject to Operator Skill-level Constraint, UTAR Institutional Repository. http://eprints.utar.edu.my/4832/1/fyp_IE_SSX_2022.pdf. Accessed April 2023.

Siltala, N., Järvenpää, E. and Lanz, M. 2018. Value proposition of a resource description concept in a production automation domain. Procedia CIRP, 72, pp. 1106–1111.

Sobek, D.K. and Jimmerson, C. 2004. A3 reports: A tool for process improvement and organizational transformation. In Proceedings of the Industrial Engineering Research Conference, Houston, Texas. 16–18 May 2004.

Sojka, V. and Lepšík, P. 2020. Use of TRIZ, and TRIZ with other tools for process improvement: A literature review. Emerging Science Journal, 4(5), pp. 319–335.

Spreafico, C. and Russo, D. 2016. TRIZ industrial case studies: A critical survey. Procedia CIRP, 39, pp. 51–56.

Subramaniyan, M., Skoogh, A., Muhammad, A.S., Bokrantz, J., and Bekar, E.T. 2019. A prognostic algorithm to prescribe improvement measures on throughput bottlenecks. Journal of Manufacturing Systems, 53, pp. 271–281.

Tague, N.R. 2005. Quality Toolbox, 2 ed., ASQ Quality Press, Milwaukee, WI. 978-0873896399.

VDI. 2019. VDI 2887 Qualitätsmanagement der Instandhaltung: Quality Management in Maintenance, Verein Deutscher Ingenieure, Beuth Verlag GmbH. Dec. 2019.

7

Analysis and Design for Operational Availability

7.1 Introduction to System Design

7.1.1 System Design Overview

7.1.1.1 Roles of System Design

The business goals for the profitability of a manufacturing company can be decomposed into a range of specific objectives. Figure 7.1 illustrates how operational factors in manufacturing, such as high throughput, effective resource utilization, quick customer response, and low inventory, contribute to a manufacturing company's profitability. System design establishes the foundation for manufacturing operation excellence, as represented by the shaded areas in the diagram.

While previous chapters focused on improving throughput within existing systems, design limitations can often constrain improvement potentials, hindered by equipment reliability, process planning issues, and maintainability. Thus, it is crucial to recognize that throughput performance is inherently constrained by the intrinsic capabilities embedded within the system's design. Design professionals need to address throughput performance during the design phases.

Manufacturers can achieve significant throughput gains by strategically optimizing system design. Such a concept, addressing performance in addition to functionality, is referred to as "design for performance." This involves the identification and resolution of throughput bottlenecks inherent in the design and various considerations during design phases. This proactive approach aims to achieve system-wide capacity enhancement, extending beyond the scope of mere production floor enhancements.

Emphasizing system capabilities early, rather than relying solely on incremental production floor improvements, is crucial for achieving long-term throughput excellence and overall business success. Both this and the next chapters explore integrating throughput considerations during design.

Design for performance, whether for throughput, quality, etc., can be challenging because it requires a good understanding of design impacts on operational performance [Trolle et al. 2020], and the assurance procedure, technical competence, and resources [Battesini et al. 2021]. Despite the recognized importance of system design, there is significant room to address design considerations and decisions to achieve the desired operational performance for manufacturing practitioners and academic researchers [Islam et al. 2022].

7.1.1.2 Phases in System Design

The tasks of developing a manufacturing system can be divided into six phases, where the primary design tasks are conducted in Phases 2–4 [Tang 2018]. Figure 7.2 depicts the progression across these three design phases.

Manufacturing System Throughput Excellence: Analysis, Improvement, and Design, First Edition. Herman Tang.
© 2024 John Wiley & Sons, Inc. Published 2024 by John Wiley & Sons, Inc.
Companion website: www.wiley.com/go/Tang/ManufacturingSystem

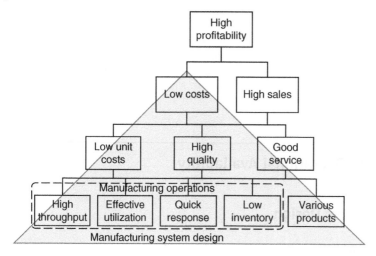

Figure 7.1   Role of manufacturing system design supporting business.

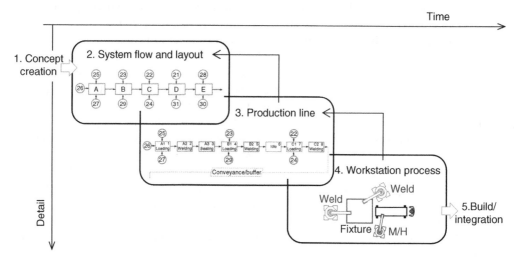

Figure 7.2   Three main phases of manufacturing system design.

1. Concept and requirement creation
2. System flow and layout design
3. Production line design
4. Workstation process planning
5. Build and integration
6. Tests and tryouts

All three design phases involve intricate tasks aimed at designing and refining various aspects. In terms of design for performance, Phases 2 and 3 play even more important roles in throughput performance and other capacities. As the primary focus here is on the key considerations for system throughput, the detailed work in the design phases is not elaborated here.

The first design Phase 2, system flow and layout design, defines and composes the entire system, as shown in Figure 7.3 illustrating the assembly lines for a car body. The output of this phase is

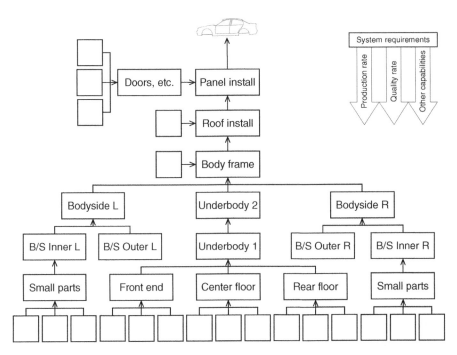

Figure 7.3 Assembly lines for vehicle bodies.

a system-level representation illustrating how the manufacturing system works, including main conveyance systems (buffers).

Following the system layout from Phase 2, the subsequent phase, assembly line design, aims to realize the system's capability requirements such as achieving a targeted throughput rate (e.g., 70 JPH) at a specified quality level (e.g., 98% first-time quality). This phase involves completing various tasks and elements, including workstation assignment, process sequencing, machinery placement, and logistical considerations, aligning with the throughput and other capability requirements. Both Phases 2 and 3 are key for addressing throughput performance. Phase 4, workstation design, continues with the process details to make all feasible, which is not the focus here as they have relatively small impacts on throughput performance.

Subsection Summary: This subsection highlights the critical role of system design for throughput in achieving profitability, underscores the importance of early integration of throughput considerations, and outlines three key design phases for throughput.

7.1.2 Discussion of System Design

7.1.2.1 Complexity of Requirements
System design involves numerous objectives, functions, constraints, and considerations, requiring the completion of pivotal tasks. This endeavor entails a series of pivotal tasks such as determining the optimal system configuration, determining the number of production lines and workstations, establishing buffer size and placement, and allocating workloads across lines and workstations.

There are numerous influencing factors and variables. They can be internal, such as budget constraints and resource availability, and external, including changes in the market and technology advancements.

Navigating complexity demands a keen focus on critical throughput-related criteria, which include the following:

- Production rate
- Quality of products
- Product unit cost
- Variety of products
- Initial investment
- Production plans and process flows
- System reconfigurability (flexibility)
- Set-up time requirement and plan
- Equipment utilization
- Manual operation requirements
- Maintenance requirements and plan
- Quantity of inventory

Most of them exhibit varying significance levels and often interact, forming a complex web of interdependencies.

To achieve the desired throughput, simplification becomes a guiding principle, minimizing the complexity of workstations when ensuring they meet specifications. For example, this involves reducing the number of process steps, minimizing tool changes, or simplifying material handling, all contributing to improved equipment utilization, reduced the number of production workers required, and simplified maintenance. Process and workstation simplification can be effectively achieved through collaborative cross-functional design work between product and process design teams, resulting in practical and streamlined design outcomes.

An innovative, simplified product design means simple manufacturing systems and processes. For example, Tesla developed its Model Y with a single rear body piece that replaced 70 distinct parts in the Model 3 [Lambert 2021]. The corresponding manufacturing systems and processes are much simpler, leading to lower costs and faster development. In addition, the size of the body assembly shop is reduced by 30%, among other benefits.

7.1.2.2 Design Task Collaboration

The design process involves various technical tasks including layout design, process planning, machinery selection, manual workstations, and logistics. These technical tasks are planned and executed with the design goals and requirements. They are successful only if the system can achieve the desired capacities and requirements.

As shown in Figure 7.4, there are the main and areas/teams cross-functional interaction in system design for throughput. The design efforts in one function area often interact with other functions due to factors associated with multiple areas. For instance, system layout design is influenced by facility design and plays a critical role in logistics planning. The three teams involved should closely coordinate their work.

Effective cross-functional teamwork is beneficial but can be challenging. Following are the three key factors merit discussion.

- Individual functional teams have their own technical requirements, budgets, timelines, and concerns. Their interests are not always the same, trade-offs can be necessary.
- The primary design goal is the entire system optimization, rather than focusing on the optimization of specific subsystems or areas. This goal necessitates effective collaboration between teams.

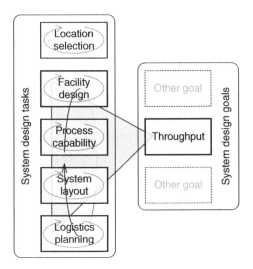

Figure 7.4 Main and cross-functional work in system design for throughput.

- From a financial perspective, the system design aims for long-term cost-effectiveness, rather than upfront cost minimization. A well-executed system design can generate increased throughput capacity, resulting in quick returns on initial investment.

This topic is extensive and closely related to project management. Senior management should encourage and enforce overall optimization, even with local technical adjustments and financial considerations within their scope. For details on these subjects, readers can refer to dedicated books. In the next chapter, throughput balanced design is explored as a technical example considering multiple factors.

7.1.2.3 Other Key Considerations

Studies have identified various considerations in system design [Alexander 1988; Barbosa and Carvalho 2013; Anderson 2020; Ginting et al. 2020]. Manufacturing system design is now considered a mature subject, with well-established processes successfully applied across industries. Numerous approaches and processes exist for manufacturing system design. Two early examples are General Motors's patent US 6931293B1 [Douglas et al. 2005] and the Massachusetts Institute of Technology's Manufacturing System Design Framework Manual [Vaughn et al. 2002].

Product and manufacturing design utilize a common set of principles and approaches, exemplified by design for quality (DFQ) [Tang 2022]. DFQ ensures that the product meets customer needs, provides delight to customers, and functions well in service. Similarly, manufacturing design professionals recognize that the customers of a system design are manufacturing plants and practitioners. Systems should be designed for operational effectiveness and efficiency, measured by various design and manufacturing KPIs. Involving manufacturing practitioners in system design is advantageous for improving system designs.

Furthermore, a significant gap exists between industry and academia in the realm of manufacturing system design. In industry, new designs typically involve modifications to existing systems, focusing on functionality with gradual and incremental changes. Consequently, there are a limited number of revolutionary innovations in production systems.

In contrast, academic researchers propose novel ideas, models, and methods, offering valuable recommendations for future design advancement. However, many of these proposals are constrained by oversimplified assumptions and models that are often far from practical implementation. It has been observed that "fewer and fewer researchers drew directly on manufacturing facilities as a source of problems" [Hopp and Spearman 2011, p. 170]. Enhancing collaboration between industry practitioners and academic researchers can bridge this gap and lead to more practical solutions.

Subsection Summary: Manufacturing system design is a multifaceted process, requiring consideration of complex requirements and various factors. Simplification and collaboration are important for achieving reliable system designs, with the aim of improving throughput and operational effectiveness. Bridging the gap between industry and academia through knowledge exchange is vital for innovative and practical solutions.

7.1.3 Design Approaches

System design involves various approaches and processes, aiming at throughput and other capabilities. This subsection provides a concise overview of the two key approaches.

7.1.3.1 V-shaped Model

As previously discussed, the development of manufacturing systems comprises the following six primary phases: concept creation, system design, production line design, process planning, build, and test. While following a sequential order, these phases exhibit significant iterative interactions. Each phase involves distinct tasks, methodologies, and tools tailored to facilitate the development process.

The six-phase system design process can be represented using a "V-shaped" model, as depicted in Figure 7.5. Tasks within each phase commonly involve iterations and encompass various aspects, including objectives, functions, and capacities.

A key advantageous feature of the V-shaped model lies in linking design activities with realization activities to ensure verification and confirmation of system functionality and capabilities, including throughput and quality. The left side of the V-shaped model covers various design activities, while the right side focuses on realizing, verifying, and validating the system's functionality and capabilities.

The model provides a systems view of better integrating of the phases in system development. As an example, consider Phase 3 on the left wing – production line design. It provides guidelines for constructing and integrating production lines, a process conducted in Phase 5. When the production line build nears completion and afterward, it undergoes testing to verify compliance with the design requirements from Phase 3. In addition, the verification process involves diverse objectives

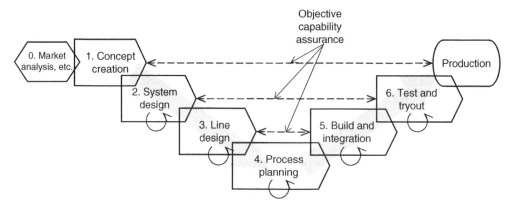

Figure 7.5 Overview of system development V-shaped model.

and focal points, including specific functionality and performance considerations [Zhou and Li 2009].

While the V-shaped model has found wide applications in various areas, such as supplier systems [Ford 2001], product design [Lubraico et al. 2003], software development [IABG 1993], and transportation systems [DOT 2007], its published documentation in the context of manufacturing system development is limited.

7.1.3.2 Proactive Design Review

Design review based on failure modes (DRBFM) process, also known as the Mizenboushi method, is a design review process developed by Toyota to proactively identify and prevent potential problems in new designs. The DRBFM process has been adopted across various fields and has become a common practice across industries since the 2000s [Thepmanee et al. 2008; Schmitt and Scharrenberg 2010; Oshima et al. 2011]. The Society of Automotive Engineers (SAE) has developed a DRBFM standard, referred to as J2886 [SAE 2013].

A DRBFM process has three steps: 1. good design, 2. good discussion, and 3. good dissection, collectively referred to as GD^3. These steps aim to achieve reliable new or modified new products and systems, as illustrated in Figure 7.6.

1. Design: The main objective of this step is to propose a new or improved design to meet requirements and be robust in a normal production environment.
2. Discussion: This step is to proactively address the potential issues and impacts of proposed design changes or new designs, based on defined objectives, functions, and capabilities. Techniques and methods, such as risk evaluation, can be used at this stage.
3. Dissection: This step involves an in-depth discussion of the countermeasures to address potential issues in the improved design. This stage may require evaluation tests to verify concerns and support countermeasures, and the outcome is an executable plan to resolve potential design issues.

During reviews, it is important to maintain a holistic, system-level perspective regarding the potential influences of specific outcomes on the entire system. For instance, modifying a manufacturing process to resolve a quality issue has no adverse impact on the cycle time of a production line. If there is an impact, e.g., lower reliability due to changing equipment, corresponding countermeasures and/or a maintenance plan should be proposed and implemented.

Subsection Summary: Two crucial design approaches are explored: the V-shaped model for manufacturing development, emphasizing the importance of system-wide considerations, and the proactive DRBFM process, a method for identifying and preventing potential issues during designs.

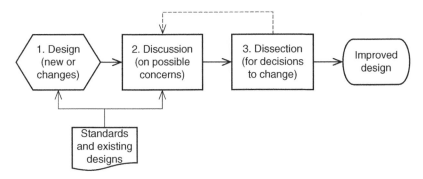

Figure 7.6 Process flow of DRBFM (GD^3) in system design.

7.2 Throughput Considerations in Design

7.2.1 Review of Throughput Rates

7.2.1.1 Different Throughput Rates

The throughput rate (TR) of a manufacturing system is determined during system design, considering various factors such as market predictions and planned production time. For example, the predicted annual market demand is 200,000 units, the production time is two shifts a day, with 7.5 hours per shift, five days per week, and 52 weeks per year. Then the total production time for a year would be $2 \times 7.5 \times 5 \times 52 = 3900$ hours. In this case, the design TR should be calculated as $\frac{200,000}{3900} = 51.28$ jobs per hour (JPH).

The design TR should be realistically achievable on the plant production floor and is often referred to as the *net* TR, representing the system's actual capacity. When designing production capacity, it is common to include a margin to accommodate uncertainties in predictions and potential future market growth. For instance, if a 25% higher TR is desired, the system's design TR would be approximately $51.28 \times 1.25 = 64.10$ JPH, using the parameters from the previous example.

The previous chapters explore how reliability affects production throughput performance, among other various factors. As discussed in Chapter 1, the gross TR is calculated under the assumptions of perfect operational reliability (or availability) and no interaction between workstations. On the contrary, the net TR of a system represents its actual throughput capacity as it considers operational availability, reflecting the throughput losses resulting from system reliability and other factors. Chapter 1 also explores the relationship between measured TR and standalone TR.

Therefore, it is essential to consider these factors when determining throughput capability during system design. As a result, there are four types of TRs for a manufacturing system. Figure 7.7 illustrates the relationship between these four types of TRs. Accordingly, the system design should ensure net TR as throughput capability. Further discussion continues in the next subsection 7.2.1.2.

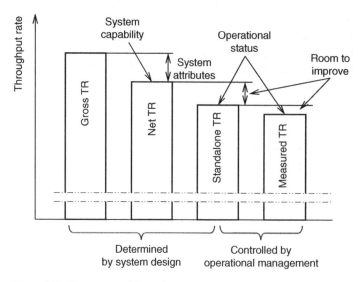

Figure 7.7 Four types of throughput rates.

7.2.1.2 Discussion of Throughput Rates

When determining the gross TR and net TR of a manufacturing system during design, it is crucial to consider key system attributes and characteristics such as reliability, cycle time, and buffers, etc. This consideration is based on various factors, such as theoretical analysis, experience from operating existing systems, computer simulations, and more. A thorough analysis can establish the foundation for ensuring optimal production performance on the production floor.

As previously discussed in Chapter 1, blockage and starvation resulting from external systems can significantly impact the throughput performance of a system, leading to the difference between measured TR and standalone TR. Production management plays a significant role in managing blockage and starvation, covered in the first six chapters.

From a system design standpoint, blockage and starvation can be minimized if they are addressed to have the balance of throughput capability across the system during design. If system design does not adequately balance the throughput capacity of the system's elements, blockage and starvation can become inherent issues in production, even though operational management tries to reduce them. Therefore, minimized blockage and starvation in design are beneficial for operational management and proactive for throughput improvement. This subject is discussed in-depth in Chapter 8.

Beyond addressing external factors like blockage and starvation, internal system design also plays a crucial role in optimizing throughput. One key aspect is quality performance, discussed in Chapter 4, significantly affecting production throughput. Ignoring quality issues or treating them separately during design leads to underestimated throughput capacity and can result in significant production losses. Using overall equipment effectiveness (OEE) is a preferred choice, including the influence of quality factor (discussed in Section 4.3), for evaluating the throughput capacity of a new system design, to be discussed in Chapter 8 as well.

Subsection Summary: Manufacturing system design should examine four distinct throughput rates (gross, net, measured, and standalone), each offering unique insights into system capability and performance. Design should consider factors such as reliability, quality, and capability balancing, for system throughput.

7.2.2 Throughput Focus in Design

Conventional system design typically emphasizes manufacturing functions and processes. However, incorporating an additional focus on throughput capacity into the traditional system design process lays a robust foundation for manufacturing operation performance on the production floor. This approach can be termed throughput-focused system design. It is important to note that throughput-focused design is not intended to supplant or replace existing design strategies or other areas of focus. Instead, it represents earlier, additional efforts aimed at addressing bottlenecks to ensure the system's capacity.

7.2.2.1 Design for Bottleneck Prevention

As thoroughly discussed in Chapter 3, addressing bottlenecks is the central focus on enhancing throughput on the production floor. Furthermore, it is vital to understand that some bottlenecks, resulting from system design, serve as the underlying causes of persistent throughput problems. These design-introduced bottlenecks are inherent to systems, posing a considerable challenge for operational management to eliminate and maintain good throughput performance over time.

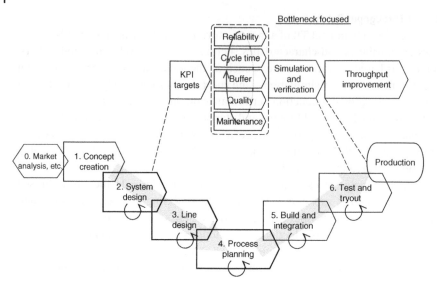

Figure 7.8 Bottleneck-free considerations in system design.

Therefore, bottleneck prevention should be a key consideration in system design. Figure 7.8 provides an overview of addressing a smooth flow of workstations without inherent bottlenecks during system design. Achieving this requires careful consideration of multiple factors (listed in the figure) that influence the system's overall performance. In addition, considering the elimination of bottlenecks, or being bottleneck-free, throughout system development.

The bottleneck-free design process involves conducting two types of feasibility reviews for each main area: technical and economic evaluations, as illustrated in Figure 7.9.

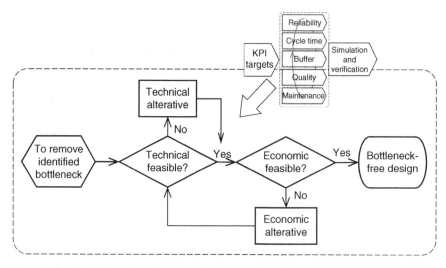

Figure 7.9 Feasibility reviews in system bottleneck-free design.

These two reviews involve a comprehensive analysis of several factors to identify and resolve potential bottlenecks. In some cases, solutions can be technically challenging or costly. The design team must explore alternative solutions or consider adjustments to other system attributes, such as maintenance or cycle time. By considering the entire system and its interdependencies, the design team can find feasible solutions for a balanced and efficient system that maximizes production output.

While achieving a completely bottleneck-free design may not always be possible, adopting a proactive approach during the design phase can significantly minimize potential bottlenecks with available resources, even within technical and financial constraints. Preventive measures for throughput (and other aspects of system performance) bottlenecks are much more cost-effective and efficient than reactive problem-solving on the production floor.

7.2.2.2 Key Considerations for Throughput Design

Manufacturing system development, inherently complex, requires meticulous consideration of numerous factors, especially system throughput performance. Here are a few key points based on industrial practices.

System design teams should cultivate proficiency in managing throughput throughout various design phases. Industrial insights underscore the need for intensified focus on throughput during the system design phase, through thorough analyses and reviews. While these efforts may require additional resources, they often result in substantial cost and time savings in the long term. This approach proves more robust than attempting to enhance throughput performance post-implementation without a dedicated emphasis on throughput in design.

In addition, the involvement of experienced manufacturing practitioners with deep knowledge of throughput and comprehensive system workstations knowledge is paramount. Their expertise can play an instrumental role in proactively averting potential throughput challenges that arise in past or existing systems, thereby aiding in preventing such issues in new systems.

Computer discrete event simulation (DES) stands out as a powerful tool in the design process, as discussed in more detail with examples in Chapter 8. However, computer simulations rely on specific assumptions and simplifications. In such situations, invaluable input and feedback from manufacturing practitioners come into play; they assist in determining the appropriate level of simplification, validating simulation outcomes, and providing insights for refining the design of the new system.

Cost considerations are pivotal in system design, involving both constraint and variability. Striving to minimize the initial investment in the manufacturing system often conflicts with the pursuit of long-term benefits arising from optimal operational efficiency. This dilemma becomes particularly pronounced when evaluating the profit generated by a manufacturing system over a specific period – a calculation contingent on profit margins and production volume. Applying the principles of throughput accounting, as discussed in Chapter 2, becomes imperative. Consequently, a comprehensive assessment encompassing elements such as initial investments, direct operating expenses, indirect costs, WIP costs, and other pertinent factors remains vital for informed decision-making in system design.

Subsection Summary: Incorporating a throughput-focused approach in system design is vital for minimizing bottlenecks, elevating operational efficiency, and optimizing system performance in manufacturing. Considering long-term costs and leveraging practitioner expertise can further enhance the likelihood of success in system design.

7.2.3 Throughput-Focused Failure Mode and Effects Analysis (FMEA)

7.2.3.1 FMEA on Throughput

FMEA is a proactive approach used to identify potential failures, assess their impacts, and recommend countermeasures during the design phases, for instance, on product, process, machinery, environment, and software. FMEA follows various standards and common practices, including those outlined in IEC [2018], SAE [2009], AIAG and VDA [2019]. Readers can find detailed information on FMEA development and evaluation ratings in sources such as Lolli et al. [2016], Korsunovs et al. [2022], and Tang [2022].

Applying the FMEA principle to system throughput during manufacturing system design can be a novel approach. A dedicated FMEA addresses throughput-related issues in design for the following purposes:

- To identify potential bottlenecks and throughput failures and address their impacts.
- To propose preventive actions and prioritize them to proactively mitigate higher risks of potential throughput issues and reduce their impacts.
- To guide design improvements to ensure improved throughput capacity, process robustness, and long-term cost reduction.

As discussed in Chapter 2, various KPIs can serve as throughput metrics. While a single-factor KPI, such as TR, can be used, a comprehensive KPI is preferred as it reflects multiple aspects of throughput performance. OEE considers the following three key throughput factors: operational availability (A), speed performance (P), and product quality (Q). From this perspective, OEE is an excellent choice as a metric for throughput FMEA development. However, failures that do not impact throughput can be addressed in separate FMEAs dedicated to other aspects of the system.

7.2.3.2 FMEA Development Process

Figure 7.10 illustrates the development steps of throughput-focused FMEA. Using OEE as the metric, there is an additional task in Steps 5 and 6 outlined in the figure, involving the review and decision on resolving a particular issue based on throughput evaluation.

1. Preparation: Forming a development team, gathering relevant data, and other necessary tasks.
2. Structure analysis: Defining the system, identifying process functions, process elements, and collecting information.
3. Function analysis: Analyzing product and process characteristics, as well as elements.
4. Failure analysis: Identifying potential failure mode, effects, and root causes.
5. Risk analysis: Analyzing and rating the Severity (S), Occurrence (O), and Detection (D) ratings of the potential failure modes. Then creating an Action Priority (AP) list or calculating the Risk Priority Number (RPN $= S \times O \times D$). If possible, estimating the corresponding OEE.
6. Optimization: Developing recommended actions based on AP or RPN, assigning responsible personnel, and implementing them. Once satisfactory AP (or RPN) and OEE (or its elements) are achieved, documentation is conducted to close the item.

Table 7.1 presents a throughput-focused FMEA form, which is very similar to other FMEA forms. The unique focus here is using OEE as an evaluation metric along with AP. Note that the latest standards recommend using AP for recommending and tracking follow-up actions, while the RPN approach is still widely used in industries. Either AP or RPN can be used.

The essence of a throughput-focused FMEA lies in assessing throughput capacity in Steps 4–6. This will be further discussed in the following subsections.

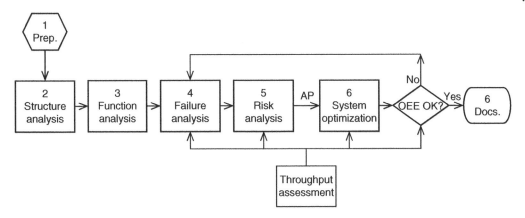

Figure 7.10 Development steps of throughput-focused FMEA.

7.2.3.3 OEE Element Ratings for FMEA

In Step 5 (Risk Analysis) and Step 6 (Optimization), OEE elements, i.e., *A*, *P*, and/or *Q*, should be assessed and rated. A dedicated rating system for OEE elements or other throughput metrics should be developed beforehand, varying depending on the specific application. Tables 7.2–7.4 provide examples of rating scales.

In these tables, the ratings are provided as a small range, such as "7–8," to allow for flexibility and judgment. Rating can be either 7 or 8 based on the element's position in the range, for example, 50–70% chance for detectability (Table 7.4) and considering other relevant factors by team consensus.

During design phases, estimating the values of impact severity, occurrence probability, and detectability may require additional studies. It is understandable that such estimates may not be entirely accurate due to limited data during early design phases. The examples provided in the tables can serve as a reference point.

Table 7.1 Throughput-focused FMEA format.

Title/ID		Area		Lead		Initial date	
Author		Team				Rev. date	

1. System	2. Process/workstation	3. Process element	1. Product/process	2. Process characteristics	3. Process element	1. Failure effect	2. Failure mode	3. Failure cause	1. Severity (S)	2. Occurrence (O)	3. Detection (D)	4. Action priority (AP), and **OEE**	1. Suggested action	2. Responsibility and target data	3. Action taken, date	4. Updated S, O, D, AP, and **OEE**
	2 Structure analysis			3 Function analysis			4 Failure analysis			5 Risk analysis				6 Optimization		

Table 7.2 Impact severity (S) rating for throughput-focused FMEA.

S	Impact	A	P	Q
9–10	Very high	<90%	<85%	<95%
7–8	High	90–94%	85–91%	95–97%
5–6	Moderate	94–97%	91–96%	97–98%
3–4	Low	97–99%	96–99%	98–99%
1–2	Very low	>99%	>99%	>99%

Table 7.3 Occurrence (O) probability rating for throughput-focused FMEA.

O	Probability		Quantitative examples	
9–10	Very high	Every 1–2 hours	~Every 100 units	>10% in a period
7–8	High	~Every shift	~Every 500 units	3–10% in a period
5–6	Moderate	~Every week	~Every 2500 units	1–3% in a period
3–4	Low	~Every month	~Every 10,000 units	0.5–1% in a period
1–2	Very low	Every several months	~ Every 20,000 units	<0.5% in a period

Table 7.4 Detectability (D) rating for throughput-focused FMEA.

D	Detectability	Quantitative example	Qualitative examples	
9–10	Very low	<50% chance	Only by a special test	Primarily by customers
7–8	Low	50–70% chance	By a periodic inspection	On main assemblies
5–6	Moderate	70–80% chance	By an in-line inspection	At the end of a line
3–4	High	80–95% chance	By a real-time inspection	At the end of a process
1–2	Very high	95–100% chance	Immediately/always	During a process

7.2.3.4 Example and Discussion

To illustrate OEE assessment in a throughput-focused FMEA, Table 7.5 shows an example.

The additional efforts, compared with conventional FMEAs for process and machinery, are the evaluation of OEE elements in Steps 4–6. For instance, Take Station 3 is analyzed in Step 4 (Failure Analysis):

1. Failure effect: Identifying possible slow operation that affects the speed performance (P) of OEE.
2. Failure mode: Specifying the nature of the failure involves a slow cycle of a robot.
3. Failure cause: the reason is identified as robot overload, a design issue.

Then, in Step 5 (Risk Analysis), the S, O, and D ratings for the throughput failures are determined using the provided rating tables. The corresponding AP is determined at a medium level, and OEE's speed performance (P) is not satisfactory.

Table 7.5 Example of throughput-focused FMEA.

2. Structure analysis		3. Function analysis	4. Failure analysis			5. Risk analysis				6. Optimization			
1. System 2. Process/workstation 3. Process element		1. Product/process 2. Process characteristics 3. Process element	1. Failure effect	2. Failure mode	3. Failure cause	1. Severity (S)	2. Occurrence (O)	3. Detection(D)	4. Action priority (AP), and **OEE**	1. Suggested action	2. Responsibility and target data	3. Action taken, date	4. Updated S, O, D, AP, and **OEE**
Station 3		Low production rate (**P**)	Slow cycle by 5%	Robot overload	5/10	8/10	2/10	Medium, **P**: NOK	Revise process	Changes made, 11/30	5/10, 2/10, 2/10, Low, **P**: OK		
Station 5		Defect/scrap (**Q**)	Screw missing (1%)	No assurance in place	8/10	6/10	8/10	High, **Q**: NOK	Need error proofing	Present sensors added (0%)	8/10, 1/10, 1/10, Low, **Q**: OK		
Station 10		Downtime (**A**)	The joining process broke (down five minutes)	Complex process	8/10	4/10	1/10	Medium, **A**: NOK	Need backup process	Manual backup added (down one minute)	5/10, 4/10, 1/10, Low, **A**: OK		

Similarly, Stations 5 and 10 can be assessed in the same manner, with Station 5 being related to OEE element product quality (Q) and Station 10 associated with operational availability (A). Following the common FMEA process, the issues are reevaluated after solutions in Step 6 (Optimization), as shown in Table 7.5.

The satisfaction criteria of OEE elements are based on throughput requirements, risk analysis, and resource constraints. In addition, throughput-focused FMEAs can be a valuable reference on the production floor.

Subsection Summary: A novel approach, using FMEA to enhance throughput in manufacturing system design, is introduced. An example of throughput-focused FMEA is presented with OEE as a metric, supported with a reference rating. This approach is adaptable to other throughput KPIs.

7.3 Reliability-Based System Design

7.3.1 System Reliability

7.3.1.1 Review of Reliability Impact

Building on the discussion in Chapter 1, a manufacturing system can stop working due to any of the four reasons: faulted, starvation, blockage, or scheduled downtime (Figure 1.19). System design and operational management have different primary responsibilities and focuses for addressing throughput issues and solutions, as illustrated in Figure 7.11. This and the upcoming chapters focus on the system design side.

A conventional understanding is that enhancing production throughput is a primary responsibility and task of operational management. However, the more effective approach to improving throughput is to integrate reliability (or availability) into manufacturing systems by design. This built-in reliability is a robust foundation for throughput enhancement by operational management on the production floor.

Reliability-centered maintenance (RCM), as discussed in Chapter 5, is an excellent approach for proactively addressing equipment reliability. The RCM concept can be extended to manufacturing systems, including appropriate maintainability, including quick inspections, accurate fault detection, easy access for repairs, and simple replacement of failed or risky parts, as all these factors directly affect MTTR.

In addition, another critical throughput enabler is that all system elements are designed for smooth operation, without major inherent blockage and starvation. From a system design

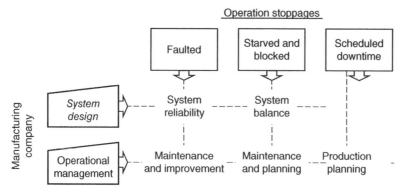

Figure 7.11 System design focuses on throughput (minimizing stoppages).

perspective, it is crucial to maintain the balance in terms of reliability across all systems and their workstations, allowing the system to operate synchronously from start to finish. Balancing throughput capacity in system design is not overly complex, but it requires careful consideration of several factors, to be discussed in the next chapter.

7.3.1.2 Reliability Analysis of System Elements

At a workstation level, multiple process steps are typically performed simultaneously; Figure 7.12 provides two examples of a similar scenario. The first example (Figure 7.12a) shows the collaboration of six robots, and the second example (Figure 7.12b) depicts a team of four operators at a single workstation. Today, it has become common for robots, referred to as collaborative robots or "cobots," to operate alongside humans in the same working environment.

A functioning workstation means all process tasks, regardless of their sequence, are conducting their assignments. A workstation's reliability (R) is the product of the individual reliability (R_i) of all (n) process elements of the workstation. This can be represented as:

$$R = R_1 \times R_2 \times \cdots \times R_n = \prod_{i=1}^{n} R_i$$

This calculation highlights that system reliability is always lower than that of any individual process within it, as each R_i is less than one.

Unlike a workstation, a manufacturing system is more complex due to its different configurations and intricate interactions among its subsystems. Chapter 1 introduced three basic layouts (serial, parallel, and hybrid), as illustrated in Figure 1.14. While a manufacturing system is overall serial, locally, there are various combinations at the levels of subsystems. Therefore, the system reliability characteristics must be evaluated accordingly.

Figure 7.13 depicts two variant examples of manufacturing design. Figure 7.13a is a U-shaped configuration of workstations allowing for efficient use of the workforce, which can be classified as a serial line. Figure 7.13b depicts a continuous-moving line with a conveyor, which can be divided into working zones, also classifying it as a serial line.

The reliability concepts discussed rely on certain assumptions, such as constant failure rates and independence among workstations, to simplify the analysis. However, these assumptions do not always hold true in real-world situations, which can affect the accuracy of analytical results. Therefore, when conducting statistical analysis, one should keep in mind that these concepts are used to gain a qualitative understanding of the system's reliability.

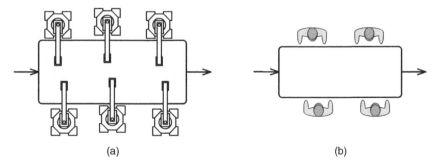

(a)　　　　　　　　　　　　　　　　(b)

Figure 7.12 Examples of manufacturing workstation designs.

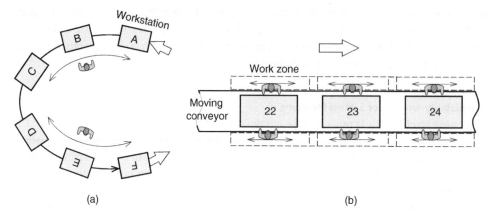

(a)   (b)

Figure 7.13   Variations of manufacturing system configurations.

Subsection Summary: This subsection highlights the significance of reliability design within manufacturing systems across various situations and analyses. These basic considerations serve as a foundation for the further reliability analysis discussed in the following subsections.

7.3.2   Reliability of Serial Systems

7.3.2.1   Theoretical Reliability Calculation

As discussed earlier, workstations are typically arranged in a serial configuration, as illustrated in Figure 7.14.

If the reliability values of all (n) workstations in a serial system are known, one can use the equation ($R_{system} = R_1 \times R_2 \times \cdots \times R_n$) from the previous subsection 7.3.1.2 to calculate its *theoretical* reliability of the system. For instance, if a system has ten workstations ($n = 10$), and each workstation has the same reliability of 99%, the system's reliability would be $0.99^{10} = 0.9044$ or 90.44%.

A system's reliability depends on the individual reliability (R_i) of all workstations and the number (n) of workstations in the system. Figure 7.15 illustrates the relationship with three cases: individual workstation reliabilities of 0.986, 0.990, and 0.994, respectively.

Manufacturing systems need to meet minimum reliability standards, such as a requirement of 90% reliability. Then, it can accommodate up to 7, 10, and 17 workstations, respectively, as shown in Figure 7.15. In other words, for a long system, its individual workstations must be very reliable.

Since individual reliability values differ, the relationship between system reliability and the number of workstations is not a straight line but exhibits fluctuations, as shown in Figure 7.16. While the overall relationship trend stays. According to the requirements of system reliability, the number of workstations can be determined for specific cases.

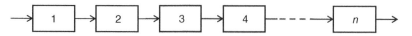

Figure 7.14   Serial-configured system with (n) workstations.

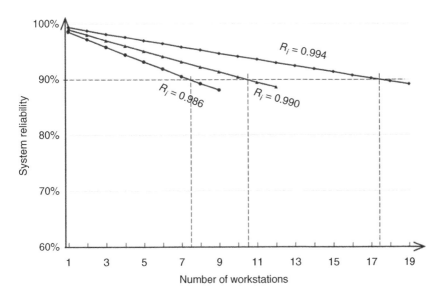

Figure 7.15 Number of workstations and resultant system reliability.

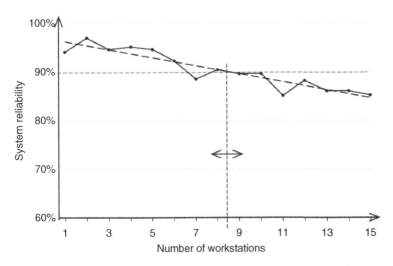

Figure 7.16 Individual workstation reliabilities and overall system reliability.

7.3.2.2 Impact of Workstation Interactions

The theoretical reliability analysis discussed is accurate under the following assumptions:

1. Each workstation is independent, meaning the failure of one workstation does not affect the failure probability of other workstations.
2. The failure of any workstation leads to the immediate failure of the entire system.
3. The failure rates for the workstations are constant.
4. The workloads on workstations are deterministic and constant in time.

A piece of equipment often satisfies these assumptions. However, for a system operational state, the first assumption of independence does not hold true. When a workstation is down, the upstream and downstream workstations are affected, either being blocked or starved. Figure 7.17 illustrates

98% reliability of each workstation				99% reliability of each workstation				100% reliability of each workstation			
	Waiting %	Working %	Blocked %	Stopped %	Waiting %	Working %	Blocked %	Stopped %	Waiting %	Working %	Blocked %	Stopped %

	Waiting %	Working %	Blocked %	Stopped %		Waiting %	Working %	Blocked %	Stopped %		Waiting %	Working %	Blocked %	Stopped %
Sta 1	0	90.983	7.04	1.977	Sta 1	0	95.15	3.856	0.995	Sta 1	0	100	0	0
Sta 2	1.206	90.982	5.791	2.021	Sta 2	0.776	95.15	3.095	0.979	Sta 2	0	100	0	0
Sta 3	2.176	90.982	4.829	2.012	Sta 3	1.439	95.15	2.489	0.922	Sta 3	0	100	0	0
Sta 4	3.067	90.982	4.027	1.925	Sta 4	1.939	87.527	9.532	1.001	Sta 4	0	100	0	0
Sta 5	3.827	90.982	3.147	2.045	Sta 5	1.758	95.15	2.121	0.971	Sta 5	0	100	0	0
Sta 6	4.721	90.982	2.277	2.02	Sta 6	2.388	95.15	1.439	1.023	Sta 6	0	100	0	0
Sta 7	5.672	90.982	1.329	2.017	Sta 7	3.093	95.15	0.79	0.967	Sta 7	0	100	0	0
Sta 8	7.036	90.982	0	1.983	Sta 8	3.851	95.15	0	0.999	Sta 8	0	100	0	0

Figure 7.17 Examples of workstation starved and blocked times.

an example of the workstation blocked or starved (waiting) time in an eight-workstation system based on computer DES. The three situations depicted in the figure correspond to 98%, 99%, and 100% reliability of each workstation, respectively. This example shows that there is no blocked or starved time only if all workstations are perfectly reliable.

Similarly, the second assumption may be argued as well. For instance, if the first workstation fails, some of the downstream workstations continue to work, delaying the shutdown of the entire system. In addition, as all the workstations are not fully synchronized, a small portion of downtime of the first workstation can be absorbed by other workstations.

In addition, there are always variations in manufacturing, including reliability, cycle time, workload, etc. Such variations show that assumptions 3 and 4 are not accurate for many situations.

Therefore, the main concern is that any workstation downtime is random and affects other workstations. The theoretical reliability calculation does not include these considerations, making the calculated results deviate from the actual situations. Although theoretical reliability can still be used for conceptual and trend analyses, more complex analyses and computer simulations are needed for the analysis of real-world systems [Negahban and Smith 2014].

7.3.2.3 Location of Less-reliable Workstation

In the context of real-world manufacturing, it is common for workstations to exhibit varying reliability. Therefore, the next question is whether the locations of less reliable workstations in a system affect its throughput. In the theoretical reliability calculation, the system reliability is only the product of all workstation reliabilities, not affected by their locations.

For discussion purposes, it is assumed that only one workstation has lower reliability and that its location can be anywhere in a system. For instance, in an eight-workstation system with a cycle time of 60 seconds, seven workstations are 99% reliable; one workstation is 95% reliable, lower than others.

Computer DES is an effective tool for estimating the actual performance of a manufacturing system, as it closely replicates real workstations, including workstation interactions. Therefore, a

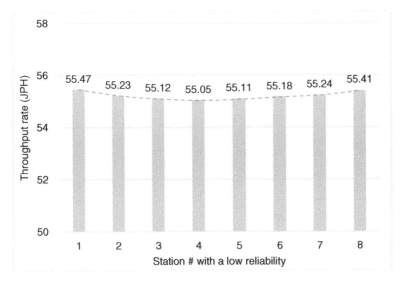

Figure 7.18 Influence of an unreliable workstation at various locations on throughput.

simulation study is conducted on the discussion example. The simulation study was conducted eight times, each time placing the less reliable workstation on one of the eight different locations. Figure 7.18 illustrates the results.

Several observations can be made from this comparison: The impact on throughput performance is not uniform, which disagrees with theoretical calculations. In this particular case, when the less reliable workstation is positioned in the center, the different influence is approximately 0.7% compared with when it is positioned at the beginning or the end. This implies other factors, including workstation interactions, are involved.

It should be noted that this example offers a comparison under simplified conditions. As mentioned earlier, workstation reliability, cycle time, and other parameters can vary. Thus, the actual effects need a DES model matching a real system to be designed.

Subsection Summary: This subsection investigates the theoretical system reliability of serial manufacturing systems, focusing on workstation interactions and their impact on performance. Through computer simulations, the subsection demonstrates the intricate relationship between reliability, interactions, and their collective influence on system throughput.

7.3.3   Reliability of Parallel Systems

7.3.3.1   Analysis for Two Parallel Segments

Expanding on the understanding of reliability analysis for serial systems, one can analyze the reliability of other configurations, for instance, parallel configurations. As this discussion focuses on concepts and principles, basic theoretical reliability is employed.

For simplicity in discussion, Figure 7.19 provides a system consisting of two serial segments in parallel. It is assumed a reliability of 0.985 for each workstation. The theoretical reliability of each segment (R_A and R_B) can be calculated individually, following the same approach used for a serial system. In this case, the theoretical reliability of each segment is $R_A = R_B = 0.985^5 \approx 92.72\%$.

To analyze the reliability of the entire system, three situations should be considered, as summarized in Table 7.6. Situation 1 presents that both segments are working, while Situations 2a and 2b present that only one segment A or B is working. Situation 3 is a state where both segments fail.

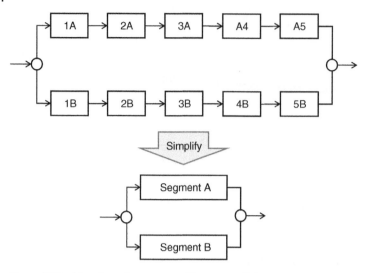

Figure 7.19 Manufacturing system with two parallel segments.

Table 7.6 Operational reliability of a parallel system.

Situation	Segment A		Segment B		States	Probability
1	Working	R_A	Working	R_B	Full working	$R_1 = R_A \times R_B = 85.97\%$
2a	Working	R_A	Failed	$1 - R_B$	Partial working	$R_{2a} = R_A \times (1 - R_B) = 6.75\%$
2b	Failed	$1 - R_A$	Working	R_B	Partial working	$R_{2b} = (1 - R_A) \times R_B = 6.75\%$
3	Failed	$1 - R_A$	Failed	$1 - R_B$	Failed	$R_3 = (1 - R_A) \times (1 - R_B) = 0.53\%$

Therefore, a system in a parallel configuration has an additional operational state compared to those in a serial configuration: one segment remains operational, or the system is partially working, in this case. The corresponding reliability for partial working is $6.75\% \times 2$, resulting in 13.50%. During partial working, the system functions but is not at full capacity (half in this case). The parallel system fully stops at a very low probability (0.53%).

In summary, a two-segment parallel system has three states with their theoretical reliability values:

1. Full working probability: $R_1 = R_A \times R_B$
2. Partial working probability: $R_2 = R_A \times (1 - R_B) + (1 - R_A) \times R_B$
3. Failure probability: $R_3 = (1 - R_A) \times (1 - R_B)$

This analysis provides insights into unique reliability characteristics, affecting system throughput in a parallel configuration.

7.3.3.2 Analysis for Multi-parallel Systems

Systems with a higher number of parallel segments present increased complexity in analysis. One can analyze them following the same process of the two-parallel systems. Here is an example of a

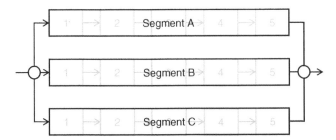

Figure 7.20 Manufacturing system with three parallel segments.

three-segment parallel system, as shown in Figure 7.20. The reliability of each segment is known as R_A, R_B, and R_C, respectively. In this case, there are four states with their theoretical reliability:

1. Full working probability: $R_1 = R_A \times R_B \times R_C$
2. Partial working probability of two segments: $R_2 = R_A \times R_B \times (1-R_C) + R_A \times (1-R_B) \times R_C + (1-R_A) \times R_B \times R_C$
3. Partial working probability of one segment: $R_3 = R_A \times (1-R_B) \times (1-R_C) + (1-R_A) \times R_B \times (1-R_C) + (1-R_A) \times (1-R_B) \times R_C$
4. Failure probability: $R_4 = (1-R_A) \times (1-R_B) \times (1-R_C)$

Like the analysis of the two-segment example before, if each segment has five workstations with a reliability of 0.985 each, then, $R_A = R_B = R_C = 0.985^5 \approx 92.72\%$ in this case. Accordingly, the system reliability of these four states can be calculated as: 79.72%, 18.77%, 1.47%, and 0.04%, respectively. The chance of all four segments failing at the same time is extremely low, at 0.04%.

An interesting finding from the reliability analysis is that if the full throughput capacity is a priority, parallel systems have the same reliability as serial systems with the same number of work-stations. From this perspective, system configurations (serial, parallel, or hybrid) do not affect the reliability of a full working state. However, there are situations where maintaining functionality, even in a partial-working state, is crucial, parallel configurations offer notable advantages over serial configurations. Again, this conceptual analysis is based only on reliability and does not con-sider interactions and other technical factors.

Furthermore, for increased flexibility, a parallel system with identical segments can be designed to have interconnections enabling multiple passes across segments. Figure 7.21a illustrates an example of a multiple-pass design; Figure 7.21b demonstrates the equivalent process flows. This can enhance the process control capability to deal with partial downtimes. The reliability of such a multiple-pass system can be evaluated in the same way.

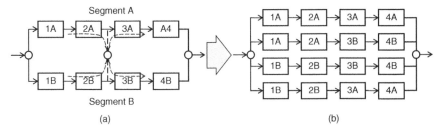

Figure 7.21 Multiple process passes with inter-segment connections.

7.3.3.3 Analysis for Hybrid Configurations

Systems can be designed with combinations of serial and parallel segments, called hybrid configurations. Figure 7.22 illustrates an example of a system with two parallel segments A and B, followed by a serial segment C. If each workstation has a reliability of 0.985, then the reliability values of segments A–C can be calculated as: $R_A = 95.57\%$, $R_B = 95.57\%$, and $R_C = 94.13\%$, respectively.

In this case, there are three states, listed and summarized in Table 7.7, with their theoretical reliability:

1. Full working probability: $R_1 = R_A \times R_B \times R_C$
2. Partial working probability of two segments: $R_2 = R_A \times (1-R_B) \times R_C + (1-R_A) \times R_B \times R_C$. In these situations, either Segment A or B works with Segment C (Situations 2a and 2b in the table).
3. Failure probability: $R_3 = (1-R_A) \times (1-R_B) \times R_C + R_A \times R_B \times (1-R_C) + R_A \times (1-R_B) \times (1-R_C) + (1-R_A) \times R_B \times (1-R_C) + (1-R_A) \times (1-R_B) \times (1-R_C)$. This situation includes five combinations (Situations 3a–3e in the table).

There can be numerous hybrid configurations such as a serial segment followed by parallel segments, two processes merging into one process, one process splitting off two or more processes.

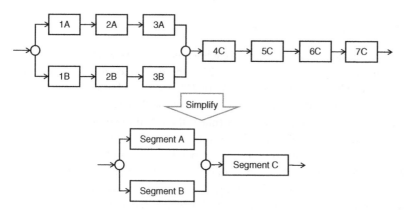

Figure 7.22 Example of a hybrid system with parallel and serial segments.

Table 7.7 Operational reliability of a hybrid system.

Situation	Segment A		Segment B		Segment C		System	Probability
1	Working	R_A	Working	R_B	Working	R_C	Full working	$R_1 = 85.97\%$
2a	Working	R_A	Failed	$1-R_B$	Working	R_C	Partial working	$R_{2a} = 3.99\%$
2b	Failed	$1-R_A$	Working	R_B	Working	R_C	Partial working	$R_{2b} = 3.99\%$
3a	Failed	$1-R_A$	Failed	$1-R_B$	Working	R_C	Failed	$R_{3a} = 0.18\%$
3b	Working	R_A	Working	R_B	Failed	$1-R_C$	Failed	$R_{3b} = 5.36\%$
3c	Working	R_A	Failed	$1-R_B$	Failed	$1-R_C$	Failed	$R_{3c} = 0.25\%$
3d	Failed	$1-R_A$	Working	R_B	Failed	$1-R_C$	Failed	$R_{3d} = 0.25\%$
3e	Failed	$1-R_A$	Failed	$1-R_B$	Failed	$1-R_C$	Failed	$R_{3e} = 0.01\%$

A typical way to model them is to divide them down into smaller, simpler segments and analyze them individually, and then integrate them for all combinations of working states.

As previously emphasized, this type of calculation will yield approximate results because the internal interactions and variations are not included. Analyzing large, complex manufacturing systems is not friendly to manual calculation. More modeling and algorithms are still under study [Gao et al. 2019].

Subsection Summary: This subsection delves into the reliability analysis of parallel systems, including two-segment and three-segment configurations, as well as hybrid configurations. It introduces the analysis process and sheds light on the crucial roles of different operational states in determining system reliability.

7.4 Additional Considerations in System Reliability

7.4.1 Relative Reliability Importance

The workstations or components of a manufacturing system have different impacts on its functions and performance. Workstations with lower reliability levels exert a more considerable influence on the system's overall reliability. To measure the relative importance of individual workstations, reliability importance (I_j) is introduced as a ratio of the system reliability (R_{sys}) over the workstation reliability $(R_j, R_j \neq 0)$:

$$I_j = \frac{R_{sys}}{R_j}.$$

Mathematically, I_j is equal to the system reliability without that workstation j. The higher the ratio, the more important the workstation is to system reliability.

To illustrate this concept, Figure 7.23 shows an example of a system with six subsystems. With known individual workstation reliabilities, shown in the figure, the theoretical system reliability is:

$$R_{sys} = 0.985 \times 0.995 \times 0.965 \times 0.985 \times 0.990 \times 0.980 = 0.904$$

The relative importance levels of workstations are calculated and listed in the third column of Table 7.8. The table also includes the rank of each workstation, calculated using Excel's function: "Rank()." In this case, the reliability importance I_3 of 0.937 is the highest, indicating Workstation 3 is the most critical component for system uptime.

For the workstations or subsystems in parallel or hybrid configurations, one can follow the approach of breaking the system into serial segments first, as discussed in the previous section. Then, find the reliability importance of segments.

For system design, the information of relative importance and rank provides a useful reference when addressing reliability concerns. In addition, one should note that the reliability of workstations can change over time depending on how they are used and maintained. It is worth additional efforts to study how the reliability of some workstations in a manufacturing system varies over time and thus changes their reliability importance.

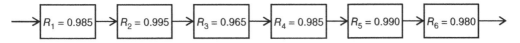

Figure 7.23 Example of a serial system with six workstations.

Table 7.8 Reliability importance ranking of six workstations of a system.

Workstation (j)	Reliability (R_j)	Importance (I_j)	Rank
1	0.985	0.918	3
2	0.995	0.908	6
3	0.965	0.937	1
4	0.985	0.918	3
5	0.990	0.913	5
6	0.980	0.922	2

Subsection Summary: Reliability importance is a measure of how significantly individual workstations or components affect overall system reliability. By calculating reliability importance, design engineers can identify critical components that have the most impact on system performance and prioritize efforts to improve system reliability.

7.4.2 Reliability Reinforcement

Some manufacturing workstations and processes have low reliability because of their technical characteristics or constraints. These types of workstations and processes often become bottlenecks in production. From a system design standpoint, using redundancy and/or backup is a common approach to increase operational availability despite low reliability.

7.4.2.1 Built-in Redundancy

Built-in redundancy in system design effectively increases the operational availability of processes or equipment. Figure 7.24 illustrates an example of adding Workstation 3B parallel to Workstation 3 to improve reliability.

The reliability of Workstation 3 with redundancy in the figure can be quantitatively evaluated. Assume Workstation 3 has a reliability of 75%, all other workstations in the system have a reliability of 95% or higher. Workstation 3 is more likely to fail and causes downtime, making it a system's inherent bottleneck. Based on the discussion of parallel system reliability in subsection 7.3.3, the reliability of Workstations 3 and 3B is $1 - (1-0.75)^2 = 93.75\%$, which is much better now.

If the reliability of 93.75% is still unsatisfactory because of other workstations with 95% reliability, additional redundancy, i.e., Workstation 3C, may be added to the system. Subsequently, the resulting reliability of Workstation 3 with 3B and 3C would be $1 - (1-0.75)^3 = 98.44\%$.

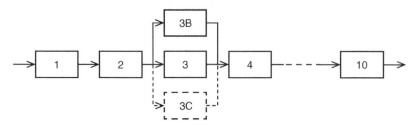

Figure 7.24 Redundancy design for an unreliable operation.

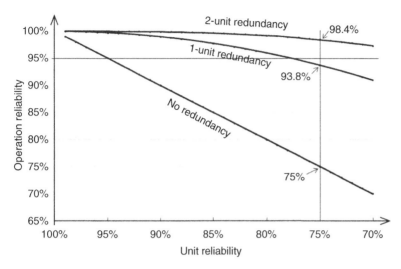

Figure 7.25 Reliability comparison between with and without redundancy.

To illustrate the impact of redundancy further, Figure 7.25 shows a general comparison of theoretical reliabilities of one unit (workstation or equipment), with single redundancy, and with double redundancy.

With redundant units, a system needs a control function that monitors the availability status of the workstations or subsystems and automatically adjusts the path accordingly without human intervention. System design should also consider the reliability of such a control system as it is not 100%.

Implementing redundancy incurs additional costs. Throughput accounting or economic analysis, considering upfront investments and the long-term benefits of reliable operations supported by redundancy, can help in decision-making, refer to the related subsections 4.2.3, 5.4.2, and 8.2.2. In general, redundancy is a worthwhile investment in high-volume production and for high-value products.

7.4.2.2 Manual Backup

In situations where unreliable or critical workstations pose a risk, implementing backups provides an effective solution when failures occur. A manual backup for an automated operation is illustrated in Figure 7.26. In this case, a human worker can intervene and perform the same functions as the automated Workstation 7 when it fails. In normal situations, the process operates without the backup, bypassing the optional backup station.

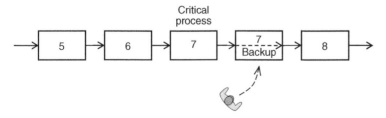

Figure 7.26 Manual backup for an automated operation.

Key considerations in designing manual backup include space, tools, and worker safety. Due to various constraints, manual backup may be slower than regular automated operations, but it can be acceptable as a temporary solution in most cases.

The throughput rate (TR) of a workstation with a slow backup can be estimated using the following equation:

$$TR_{w\ backup} = TR_{gross} \times R + TR_{backup} \times (1 - R)$$

where R is the reliability of the workstation, and TR_{backup} is the TR of the backup operation.

For instance, a workstation has a reliability of 80%, with a gross TR of 60 JPH, and a backup with a TR of 40 JPH. Then the estimated TR of the workstation during backup would be:

$$TR_{w\ backup} = 60 \times 0.8 + 40 \times 0.2 = 56\ (JPH)$$

In the equation, it is assumed the backup is fully reliable. If the backup is not fully reliable, then the second item in the equation should multiply the backup reliability (R_{backup}) as:

$$TR_{w\ backup} = TR_{gross} \times R + TR_{backup} \times (1 - R) \times R_{backup}.$$

7.4.2.3 Automated Backup

An automated backup, not a permanent redundancy, can be a better option to improve reliability and throughput if financially justifiable. Figure 7.27a shows a scenario of automatic backup in the system. In this example, Robot 7 can substitute Robots 1, 3, or 5 when needed. Similarly, Robot 8 can take over the work of any robot on the opposite side. This automated backup design can maintain full speed and functionality in case of a robotic failure.

A partial backup is another option, as shown in Figure 7.27b. This design uses existing equipment to cover a piece of faulty equipment. In this example, Robots 1 and 5 can assist Robot 3 when it fails to keep the production running. However, this backup design can lower the system speed because Robots 1 and 5 have more work to do when Robot 3 fails. This option increases the complexity of control logic, robot programming, and has limitations in backing up only for certain types of operations.

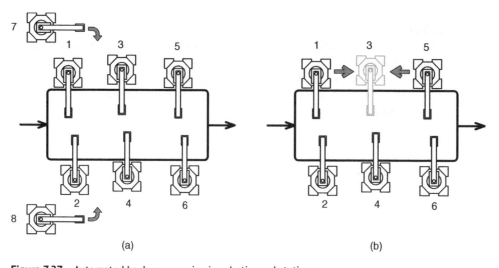

Figure 7.27 Automated backup scenarios in robotic workstations.

With automated backup ready on standby, the theoretical reliability (R) of a workstation can be significantly increased. The improved reliability with backup can be presented:

$$R_{improved} = R + [(1 - R) \times R_{backup} \times R_{switch}],$$

where R_{backup} and R_{switch} are the reliability values of the backup and switch functions, respectively. The increased reliability is approximately $[(1-R) \times R_{backup} \times R_{switch}]$. For example, if Robots 3 and 7 in Figure 7.27a have a reliability of 97%, and the backup switch has a reliability of 95%, then the reliability of Robot 3 with the backup of Robot 7 is:

$$R_{3+backup} = R_3 + (1 - R_3) \times R_7 \times R_{switch} = 0.97 + 0.03 \times 0.97 \times 0.95 \approx 0.9976.$$

Subsection Summary: This subsection explores enhancing manufacturing system reliability through three design options: built-in redundancy, manual backups, and automated backups. It also analyzes the effects of these design options, providing insights into optimizing manufacturing systems by design.

7.4.3 Validation Tests

7.4.3.1 Capability and Reliability Tests

Beyond calculations and simulations, physical validation tests are crucial to verify that a manufacturing system's design aligns with its intended functions and capabilities, including reliability. This sixth step in the system development V-model (Figure 7.5) involves rigorous testing and tryouts, each with specific objectives and passing criteria.

Multiple tests and tryouts are conducted for validation purposes, each test with corresponding specific objectives and passing criteria. For instance, evaluating fixture capability often involves assessing accuracy and repeatability as crucial factors.

Validation tests and tryouts are typically conducted in the two phases: Phase 1 assesses individual functions and capabilities, e.g., fixture accuracy and repeatability. Phase 2 evaluates the system's overall reliability and performance. Figure 7.28 illustrates an example of comprehensive testing for a vehicle body assembly line.

The passing criteria for the validation tests should be at least as stringent as the requirements in regular production. Because the validation tests usually involve smaller sample sizes and shorter durations, their results can exhibit greater variations. Therefore, many tests need to be repeated multiple times to reveal issues and ensure consistent, reliable results.

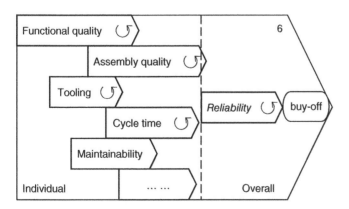

Figure 7.28 Tests and tryouts of vehicle assembly line development.

Manufacturing system reliability tests can serve as a conclusive, assuming all functions operate as intended in a production environment. Therefore, all functional tests must have been successfully completed.

A common type of manufacturing system reliability test is a dry run, performing nearly all production functions without producing an actual product. A dry run test is practical as not all product parts are available for new products when a manufacturing system is about ready. Such a dry run test should be run for a predetermined period, such as 16 hours, without experiencing any major failures.

Active participation in validation testing by the system design teams is essential, enabling prompt responses to any issues that may occur during testing, in addition to fulfilling their design and delivery responsibilities. By adhering to these structured validation processes, the system development team can confidently launch systems that consistently deliver the performance they promise.

7.4.3.2 Problem-Solving in Tests

It is uncommon for validation tests and tryouts to pass on the first attempt. They can reveal design issues and request changes to the solution before the systems are ready to be put into production. Then, later tests can be used to validate design adjustments and changes made.

The root causes of issues and failures identified in the tests can be attributed to several factors, including design, construction, setup, and debugging, among others. These root causes must be addressed accordingly and resolved before retesting. Common problem-solving approaches discussed in Chapter 6, including PDCA, DMAIC, 8D, and A3, can be effectively used to address the issues and failures.

As a part of the problem-solving process, issues, failures, and other concerns that arise during the tests should be documented on-site and reviewed immediately after the test for follow-up actions. Then, it is important to monitor the progress of corrective actions to ensure issues, failures, and concerns are addressed and resolved on time.

For facilitating this process, it is recommended to use a simple tracking form, such as the one shown in Figure 7.29. In monitoring and tracking, the current state of each issue can be

Administrative information (System/process, test, builder, date, track number, supervisor, etc.)									
Issue	Descript.	Method/spec	Sample	Date	Action req'd	Assigned	Due date	Results	Status
1	Slow cycle Sta.9	Tool, requirements standard ◊	5x	11/8/22	Reprograming robots	Mike T.	11/11/22	Resolved report ◊	Complete
Authorized/approval signature, date, and remarks									

Figure 7.29 Test follow-up tracking form in design validation.

color-categorized in the status column for easy recognition. For example, green shows that an issue is resolved, yellow indicates that the issue is being resolved, and red signals that the issue is behind schedule.

After all serious issues are resolved and the final reliability tryout is satisfactory, a formal acceptance process and documentation, often referred to as a buy-off in practice (Figure 7.28), is conducted and signed to approve the system's design, construction, and integration.

It is important to note that these validation tests and tryouts are designed to be confirmatory in nature, meaning that they solely validate the system's capability under specific circumstances (for example, limited sample size, duration, and untested functions related to physical parts). This implies that while overall capability appears satisfactory, it does not provide an absolute guarantee of all the system's capabilities.

Manufacturing systems are complex, and limited validation tests cannot uncover every potential issue within a given period. In addition, resolved problems may resurface, and new issues can arise during normal production. This underscores the vital role of production management and continuous improvement in achieving throughput excellence as discussed in earlier chapters.

Subsection Summary: This subsection briefly discusses that validation tests are crucial in system development, to confirm performance under specific conditions and ensure readiness for implementation. This process involves monitoring, problem-solving, and a formal acceptance procedure for new manufacturing systems.

Chapter Summary

7.1. Introduction to System Design

1. System design is pivotal in achieving optimal profitability for manufacturing businesses by addressing specific objectives and responsibilities.
2. Early integration of throughput considerations during design is essential to minimize limitations and optimize system performance.
3. System design focuses on production throughput and product quality, guided by KPIs such as reliability, cycle time, and maintenance effectiveness.
4. Manufacturing system development involves three major phases: system flow, production line, and workstation process.
5. Some approaches are important to system development for throughput, including the V-shaped model and DRBFM (design review based on failure mode).

7.2. Throughput Considerations in Design

6. Throughput rates are determined based on market predictions, planned production time, and realistic achievable capacity.
7. There are four types of throughput rates: gross, net, standalone, and measured.
8. Thorough analysis, experienced professional involvement, and total cost considerations are crucial in designing systems.
9. Addressing and preventing inherent bottlenecks is a key to manufacturing performance design.
10. OEE or other throughput KPI should be incorporated into the FMEA during system development to ensure throughput capacity.

7.3. Reliability-Based System Design

11. System design should integrate reliability, addressing faults, starvation, blockages, and scheduled downtime during workstations.
12. The principle of reliability-centered maintenance (RCM), introduced in Chapter 5, can extend to designing reliable manufacturing systems.
13. A system's reliability is influenced by its configuration, including serial, parallel, or hybrid, in addition to workstation reliability.
14. The actual reliability of a system often deviates from its theoretical reliability due to dynamic internal interactions and variations.
15. Reliability analysis for a parallel system is more complex than for a serial system because of its partial working status.

7.4 Additional Considerations in System Reliability

16. The importance of each workstation or equipment in a system can be assessed based on its impact on the system's reliability.
17. Inherent reliability issues can be addressed through redundancy or backup processes, with careful consideration of cost and benefits.
18. Physical tests are essential in validating manufacturing system functions, capabilities, and reliability, providing insights for design improvement.
19. Validation tests often reveal design issues, requiring problem-solving approaches to follow up.
20. A formal acceptance process is conducted after successfully passing reliability tests, marking approval for the system's design, construction, and integration.

Exercise and Review

Exercises

7.1 A manufacturing system has seven workstations in series with reliability values of 0.98, 0.95, 0.97, 0.98, 0.95, 0.98, and 0.98, respectively. Calculate the theoretical reliability of the system.

7.2 A manufacturing system has two parallel subsystems, each with five workstations with a reliability of 0.98. Determine the system's reliability when operating at full capacity (both subsystems working) and at half capacity (only one subsystem working).

7.3 A manufacturing system comprises two parallel subsystems and one serial subsystem, each with five workstations, as shown in Figure 7.30. The workstation reliability values in the

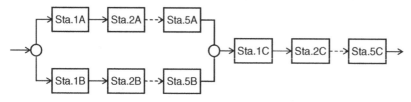

Figure 7.30   For Exercise 7.3.

parallel subsystems *A* and *B* are 0.97, and those in the serial subsystem *C* are 0.98. All subsystems have five workstations. Calculate the system's reliability when operating at full capacity (all subsystems working) and at half capacity (one parallel subsystem working with the serial subsystem). (Refer to Table 7.6 for a parallel system analysis.)

7.4 A manufacturing system has seven workstations in series with reliability values of 0.98, 0.96, 0.97, 0.98, 0.95, 0.98, and 0.99, respectively. Identify the workstation most critical to system throughput in production.

7.5 A process has a reliability of 85% and a gross TR of 80 JPH. If a slow manual backup with a TR of 60 JPH can be designed for the process to improve its reliability, what is the TR of the process with backup over an extended period? What is the TR if the manual back is only 80% reliable?

7.6 A process has a reliability of 0.85. To improve its reliability, an identical process is designed as an automated backup with a switch reliability of 0.97 and integrated into the operation. What is the improved reliability of the operation?

Review Questions

(The chapter covers these topics. For further discussion, it is recommended to seek additional information and examples. Diverse perspectives are encouraged.)

7.1 Select two system design objectives, other than throughput, and discuss their relationships with the system's throughput capability.

7.2 Select one design enabler (refer to Figure 7.2) and explain how it influences the system's throughput capability, providing an example.

7.3 Provide an example and discuss a system design objective for manufacturing.

7.4 Outline the three main phases involved in the manufacturing system development using an example.

7.5 Explain why iterating main design tasks (including facility, process, layout, and logistics) is necessary to meet throughput requirements, providing an example.

7.6 Describe how the V-shaped model can be applied in the manufacturing system design, providing an example.

7.7 Describe the DRBFM process and give an example of its use in the design of a manufacturing system.

7.8 Compare and contrast the four types of manufacturing system throughput rates. Which one(s) should be focused on during system design?

7.9 Discuss the impact of product design on the throughput capability of a manufacturing system, providing an example.

7.10 Review the technical and economic feasibility of alternatives and their integration into the system design for throughput optimization.

7.11 Discuss the concept of throughput-focused FMEA and its potential applications. Can we consider throughput rate instead of OEE?

7.12 Review the distinct impacts of system design and operational management on system throughput and discuss how they can work together.

7.13 How can the reliability values of individual workstations and the number of workstations affect the overall system reliability?

7.14 Do the interactions between workstations in a system affect the calculated system throughput based on known cycle time and reliability? Explain.

7.15 Compare the reliability of a parallel system with a similar serial system and highlight the advantages of a parallel system in terms of operational availability or throughput.

7.16 Explain the concept of partial operational availability in parallel systems and its implications.

7.17 Discuss the calculation and interpretation of a workstation's reliability importance for the overall system's reliability.

7.18 Discuss the advantages and disadvantages of workstation redundancy by design, providing an example.

7.19 Provide an example and discuss the necessity for backup functions in some manufacturing workstations during system design.

7.20 Describe the validation tests used in system design and provide an example.

References

AIAG and VDA 2019. AIAG and VDA FMEA Handbook, ISBN-13: 978-1605343679, Automotive Industry Action Group (AIAG) and German Association of the Automotive Industry (VDA), Southfield, MI.

Alexander, A.J. 1988. The Cost and Benefits of Reliability in Military Equipment, RAND Corp, Santa Monica CA.

Anderson, D.M. 2020. Design for Manufacturability: How to Use Concurrent Engineering to Rapidly Develop Low-cost, High-quality Products for Lean Production, Productivity Press.

Barbosa, G.F. and Carvalho, J.D. 2013. Design for Manufacturing and Assembly methodology applied to aircrafts aircraft design and manufacturing. IFAC Proceedings Volumes, 46(7), pp. 116–121.

Battesini, M., ten Caten, C.S. and de Jesus Pacheco, D.A. 2021. Key factors for operational performance in manufacturing systems: Conceptual model, systematic literature review and implications. Journal of Manufacturing Systems, 60, pp. 265–282.

DOT. 2007. Systems Engineering for Intelligent Transportation Systems, US Department of Transportation. https://ops.fhwa.dot.gov/publications/seitsguide/seguide.pdf. Accessed May 2023.

Douglas M., Kornas, E., Anticoli, M. et al. 2005. Method For Early Optimization of a Manufacturing System Design. U.S. Patent No. US 6,931,293 B1. U.S. Patent and Trademark Office. https://patft.uspto.gov/netacgi/nph-Parser?Sect1=PTO1&Sect2=HITOFF&p=1&u=/netahtml/PTO/srchnum.html&r=1&f=G&l=50&d=PALL&s1=6931293.PN.

Ford. 2001. Advanced Product Quality Planning (APQP) Status Reporting Guideline, Ford Motor Company, Quality Office. March 2001. https://elsmar.com/Cove_Members/APQP/APQP_Misc%20Folder/APQP_Ford_2001.pdf. Accessed May 2023.

Gao, S., Rubrico, J.I.U., Higashi, T., Kobayashi, T., Taneda K., and Ota, J. 2019. Efficient throughput analysis of production lines based on modular queues, IEEE Access, 7, pp. 95314–95326. https://doi.org/10.1109/ACCESS.2019.2928309.

Ginting, R., Ishak, A., and Malik, A.F. 2020. Product development and design with a combination of design for manufacturing or assembly and quality function deployment: A literature review. In AIP Conference Proceedings, Vol. 2217, No. 1. AIP Publishing LLC, pp. 030159.

Hopp, W.J. and Spearman, M.L. 2011. Factory Physics: Foundations of Manufacturing Management, 3 ed., ISBN: 978-1577667391, Waveland Press, Long Grove, Illinois.

IABG. 1993. V-Model Lifecycle Process Model, IABG Information Technology. http://www.v-modell.iabg.de/kurzb/vm/k_vm_e.doc. Accessed May 2023.

IEC. 2018. IEC 60812:2018 Failure Modes and Effects Analysis (FMEA and FMECA), International Electrotechnical Commission, Geneva, Switzerland.

Islam, M.H., Chavez, Z., Birkie, S.E. and Bellgran, M. 2022. Enablers in the production system design process impacting operational performance. Production & Manufacturing Research, 10(1), pp. 257–280.

Korsunovs, A., Doikin, A., Campean, F., Kabir, S., et al. 2022. Towards a model-based systems engineering approach for robotic manufacturing process modelling with automatic FMEA generation. Proceedings of the Design Society, 2, pp. 1905–1914.

Lambert, F. 2021. Tesla is Building Model Y Bodies with Single Front and Rear Castings, A Manufacturing First, electrek. https://electrek.co/2021/10/05/tesla-building-model-y-bodies-single-front-rear-castings-manufacturing-first/. Accessed March 2023.

Lolli, F., Gamberini, R., Rimini, B. and Pulga, F. 2016. A revised FMEA with application to a blow moulding process. International Journal of Quality & Reliability Management 33 (7), pp. 900–919.

Lubraico, M. et al. 2003. Vehicle Program Management Concept, SAE Paper No.2003-01-3644, SAE International, Warrendale, PA, USA.

Negahban, A. and Smith, J.S. 2014. Simulation for manufacturing system design and operation: Literature review and analysis. Journal of Manufacturing Systems, 33(2), pp. 241–261.

Oshima, M., Nara, K. and Yoshimura, T. 2011. Prevention of defects and customer dissatisfaction using quick design review. Proceedings of the Society of Automotive Engineers of Japan, 42(2), 657–662. SAE Technical Paper 2011-01-0510.

SAE. 2009. Potential Failure Mode and Effects Analysis in Design (Design FMEA), Potential Failure Mode and Effects Analysis in Manufacturing and Assembly Processes (Process FMEA), J1739, SAE International, Warrendale, PA.

SAE. 2013. Design Review Based on Failure Modes (DRBFM), J2886, SAE International, Warrendale, PA.

Schmitt, R. and Scharrenberg, C. 2010. Approach for Improved Production Process Planning by the Application of Quality Gates and DRBFM. In Proceedings of the 6th CIRP-Sponsored International Conference on Digital Enterprise Technology, Springer Berlin Heidelberg. pp. 1089–1100.

Tang, H. 2018. An integrated product-process hierarchical modeling method for development of complex assembly manufacturing systems, 7th CIRP Conference of Assembly Technologies and Systems, Procedia CIRP (ISSN: 2212-8271), Vol. 76, pp. 2–6.

Tang, H. 2022. Quality Planning and Assurance—Principles, Approaches, and Methods for Product and Service Development, ISBN-13: 978-1119819271, Wiley, Hoboken, NJ.

Thepmanee, T., Sirilappanich, S., Pongswatd, S. and Ukakimaparn, P. 2008. Application of GD 3 in value engineering for plastic-part design. In 2008 International Conference on Control, Automation and Systems, IEEE. pp. 1908–1912.

Trolle, J., Fagerström, B. and Rösiö, C. 2020. Challenges in the fuzzy front end of the production development process. In SPS2020, IOS Press. pp. 311–322.

Vaughn, A., Fernandes, P., and Shields, J. T. 2002. Manufacturing System Design Framework Manual, MIT Libraries. https://dspace.mit.edu/handle/1721.1/81902. Accessed September 2022.

Zhou, J. and Li, D. 2009. Reliability Verification: Plan, Execution, and Analysis, SAE Technical Paper 2009-01-0561. https://doi.org/10.4271/2009-01-0561.

8

System Design for Throughput Assurance

8.1 Buffer Planning for Throughput

There are numerous factors that influence system performance, including issues and variabilities related to system design and operational management, which may lead a manufacturing system to perform below expectations. As discussed in Chapter 3, buffering absorbs some of the variabilities and minor issues, enabling smoother operations and better performance. Buffers are needed due to the inevitability of variabilities [Hopp and Spearman 2021].

Buffers can take the form of WIP units, capacity, time, or a combination. This chapter examines WIP buffers, often manifested as physical conveyors in manufacturing systems. The discussion also includes capacity and time buffers, which can compensate for other capacity constraints to achieve a balanced system design. For optimizing production flow and throughput, buffers should be designed with effectiveness in mind, considering factors such as shop floor size, feasible techniques, and cost.

In addition to conveyor systems, the mechanism for resequencing or batching WIP units can serve as a WIP buffer. For example, an automated storage and retrieval system (ASRS) is implemented as a subsystem within a larger manufacturing system. Turner [2020] identified the following five key factors for evaluating ASRS design and selection: inventory capacity, throughput, modularity, maintenance, and cost. An important consideration is the tradeoff between inventory size and throughput performance in the design of an ASRS.

8.1.1 Buffer Allocation Planning

8.1.1.1 System Decoupling

As discussed in Chapter 7, long manufacturing systems are typically considered unreliable. An effective way to reduce the length of a system is to place buffers in the system, dividing it into several segments and enabling them to operate somewhat independently. Figure 8.1 provides a visual representation of dividing a 20-workstation system into three segments by placing two buffers. This approach boosts the system's operational availability without altering the reliability of workstations.

The next question is the strategic positioning of these buffers to have better cushion effects. Utilizing a computer simulation, researchers have examined various buffer positions within a system to assess their impact on performance.

Figure 8.2 illustrates an example of a serial system with 12 workstations and an internal buffer with a capacity of ten WIP units. Buffer effects, as studied through simulations, show consistent patterns across buffer sizes, workstations, cycle times, and workstation reliabilities [Imseitif and

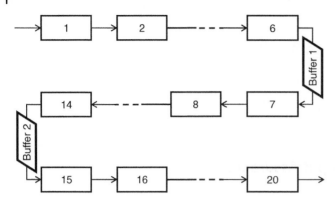

Figure 8.1 Decoupling a long system into segments with buffers.

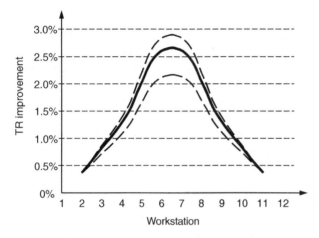

Figure 8.2 Effects of buffer locations in a system for throughput improvement.

Tang 2019a; Imseitif et al. 2019b]. For instance, smaller and larger buffer sizes correspond to smaller and larger effects on throughput, respectively, as illustrated in the dashed-line curves in the figure.

Placing a buffer at the system's halfway point, if feasible, can be a general guideline for maximizing throughput gains. It is essential to acknowledge that certain constraints, such as process and technical requirements and floor space limits, must be considered when determining buffer placements. Accordingly, the proposed buffer locations can be decided optimally and practically to enhance throughput.

As discussed in subsection 7.3.2, the required reliability of a system limits the maximum number of workstations. In such cases, the number of buffers can be determined to decouple a long system into segments.

8.1.1.2 Buffer for Variation Reduction

As discussed in Chapter 6, variation is always present in manufacturing systems and performance. Manufacturing variability can be defined as any deviation on the production floor from the designed nominal performance. Reducing variation is a key focus of continuous improvement, involving the identification and addressing of controllable or assignable root causes. The inherent

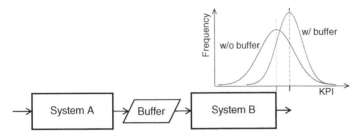

Figure 8.3   Illustration of buffer effect on variation reduction.

sources of variability, known as common causes, should be considered in system design. Buffers can reduce variation in operations (see Figure 8.3). Buffers play a critical role in reducing variation, particularly due to common causes. Most variation resulting from controllable root causes can be mitigated through good operational management.

In systems with high inherent variability, various types of buffers, such as raw material, WIP inventory, extra technical capability, and/or time, should be employed. These buffers help reduce variability and ensure a smooth production flow. This concept, known as the "variability buffering law" [Hopp and Spearman 1995], has been studied in various systems, as evidenced by examples such as Guide and Srivastava [2000] and Stratton [2008].

For instance, the incoming logistics of materials and parts may not always be on time due to the variability in transportation and delivery. Even a slight delay or disruption can affect production lines if there are no buffers for the parts. Therefore, a buffer of inbound parts for the production lines can ensure continuous operations. Similarly, inventory space should serve as a buffer at the end of the system to prevent blockage and outbound shipment delays. Subsection 8.3.1 later in this chapter discusses how time tolerance and time buffers may be used to offset variation for critical workstations. Manual operations also tend to exhibit high inherent cycle time (CT) variation, necessitating certain buffer compensation.

8.1.1.3   Buffer for Unreliable Operations

Building on Chapter 7's discussion, equal reliability of workstations is ideal and often assumed within a system. However, the reality is that different processes and equipment lead to differing workstation reliability. Certain workstations or operations have lower reliability, including complicated processes, the use of complex high-technical equipment, and a combination of automation and manual operations. Low-reliable workstations are inherent and likely bottlenecks in production operations.

Alongside efforts to improve reliability, adding a buffer before and after a low-reliable workstation can be a viable solution to reduce the likelihood of it becoming a bottleneck in production. Figure 8.4 illustrates these two scenarios of buffer placements.

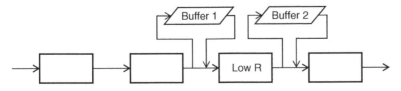

Figure 8.4   Buffers before and after an expected bottleneck operation.

Employing a front buffer and an after buffer (Buffers 1 and 2 in the figure) can notably enhance production flow and throughput by facilitating a smoother flow. In the design shown in the figure, the use of the buffers is optional when necessary. The buffer sizes can be determined through computer simulations and/or based on experience. In addition, the physical locations of buffers depend on floor space. It is a good option if buffers can be fully integrated into the line.

The determination of the need for buffers and their sizes can be achieved through computer simulations, facilitating a data-driven decision-making process. For example, adding a front buffer without an after buffer may be suitable for some cases. The buffer effect results can also be obtained by leveraging prior experience and lessons learned. For instance, historical data on the average frequency of downtime and variation in CT for similar operations can serve as valuable references for buffer sizing.

Subsection Summary: This subsection explores how buffer allocation can enhance the performance of manufacturing systems. By decoupling a long system into smaller segments, reducing variability in production output, and mitigating the impact of unreliable workstations, strategically placed buffers can lead to increased system throughput.

8.1.2 Buffer Path Design

8.1.2.1 Carrier Return Loop

Large conveyor systems, such as tooling pallet shuttles, frequently operate as closed-loop systems. They are widely used in automated manufacturing processes, such as vehicle assembly lines. A tooling pallet shuttle serves as a mobile fixture system, transporting WIP units on pallets within a manufacturing system. In a pallet shuttle loop, tooling pallets carrying WIP units progress forward along the manufacturing processes, while empty tooling pallets return to the system's starting point.

To illustrate this scenario, Figure 8.5 shows an example of a vehicle body assembly system. The assembly process begins at the starting point marked with "○" in the Underbody 1 system and

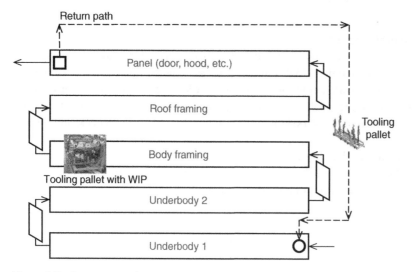

Figure 8.5   Return path of tooling pallets in a large manufacturing system.

concludes at the endpoint marked with "□" in the Panel system. Subsequently, empty tooling pallets traverse back from point "□" to point "○" along the dashed line in the figure.

In such a design, the return path itself is a conveyor and serves as a buffer, which can be analyzed accordingly. In the case of a long return path, its capacity (the maximum number of empty pallets) is high. Being half full or less can provide sufficient buffering to the beginning of the manufacturing system. However, the minimum number and transfer speed of empty tooling pallets on the return path can influence system performance, by potentially starving the system.

For instance, a tooling pallet takes 15 minutes on the return path, and the system cycle time is 45 seconds. The minimum number of units, as discussed in subsection 3.3.2 of Chapter 3, on the return path can be calculated as follows:

$$\text{Min} = \frac{t_{\text{transfer}}}{\text{CT}} + 1 = \frac{15 \times 60}{45} + 1 = 21 \text{ (units)}.$$

Operating below the minimum leads to a high probability of starvation at the beginning of the system during production. The optimal number of empty tooling pallets should be determined through computer simulation. The conveyor path is further discussed in the following subsection 8.1.2.2.

8.1.2.2 Long Path and Shortcut

Based on the understanding of the return path and its impact on system throughput, specific considerations can be examined for the enhanced efficiency of tooling pallet shuttle systems.

Several solutions exist to enhance the efficiency of a tooling pallet shuttle system. For instance, one approach involves minimizing the return path distance through thoughtful system layout design. Another straightforward option is to increase the moving speed in the return path. In addition, designing a dual path – comprising a long path and a shortcut – within a large system proves to be an effective option if space allows.

In a dual-path design, a long path provides additional WIP buffering capability, better accommodating variations in demand, downtimes, and other types of minor issues. Meanwhile, a shortcut path reduces travel time and the possibility of starvation.

Figure 8.6 displays a real-world system featuring a dual-path design. Figure 8.6a depicts the system layout, and Figure 8.6b illustrates the simplified conveyance flow. In the figure, two paths connect manufacturing System A and System B: (1) shortcut and (2) long path. Systems A and B are at the floor level, while the pallet shuttle conveyance (shaded in the figure) is elevated above the floor level.

The dual-path design has significant benefits, covering both added buffer size and improved responsiveness. While a dual-path design offers both buffer size and responsiveness, it also comes with higher initial investment and the need for a control routing system to monitor the number of WIP units on both paths and select which one to select from based on real-time production data and system status. Therefore, adopting a dual-path design is based on its cost-effectiveness in the long run.

Subsection Summary: This subsection explores strategies for designing return paths in complex tooling pallet shuttle systems. It introduces a dual-path conveyance design, which balances buffering capacity with improved responsiveness for enhanced production flow.

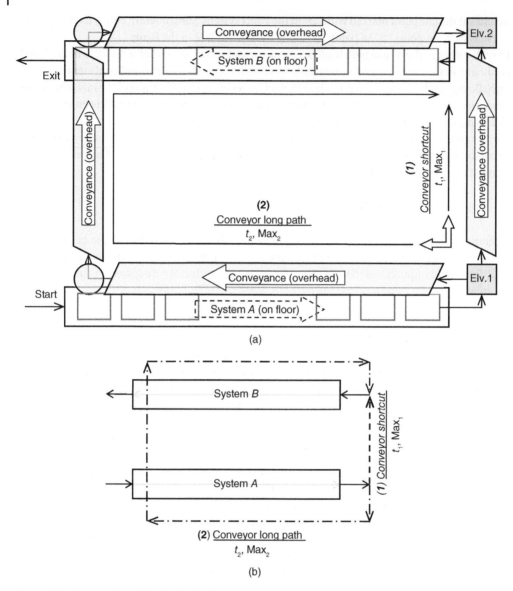

Figure 8.6 Long path and shortcut path of a conveyor system.

8.2 Analysis in Buffer Design

8.2.1 Buffer Effect Estimation

8.2.1.1 Estimation Considerations

To analyze buffer effects on system throughput performance, two system designs can be compared: without and with a buffer, as illustrated in Figure 8.7a and Figure 8.7b, respectively.

In the no-buffer design, the theoretical net throughput rate (TR_{net}) of the entire system can be approximated by:

$$TR_{net} = Min\{TR_1 \times A_1, TR_2 \times A_2\},$$

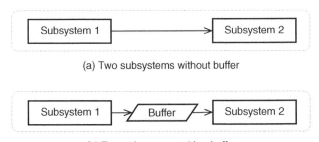

(a) Two subsystems without buffer

(b) Two subsystems with a buffer

Figure 8.7 System designs with and without a buffer.

Table 8.1 Capability parameters of a sample system.

Subsystem	Capability	
1	$TR_1 = 59$ JPH	$A_1 = 94.5\%$
2	$TR_2 = 60$ JPH	$A_2 = 93.5\%$

where TR_1 and TR_2 are the gross throughput rates (without downtime or reliability issues) of Subsystems 1 and 2, respectively. A_1 and A_2 are the operational availability of Subsystems 1 and 2, respectively. Min{} represents the minimum function to select the smallest number within {}.

The values of TR and A are typically known or can be estimated. Table 8.1 lists them for the two subsystems as a discussion example.

Using the formula, one can calculate the TR_{net} of the design without a buffer:

$$TR_{net} = Min\{59 \times 94.5\%, 60 \times 93.5\%\} \approx Min\{55.8, 56.1\} = 55.8 \text{ (JPH)}.$$

As discussed in Chapter 3, a buffer between two subsystems can reduce the blocked time for Subsystem 1 and the starved time for Subsystem 2, improving system throughput. However, the buffer effect is dynamic and related to real-time operational status. Therefore, it can only be estimated using statistical principles, assuming stable operations without a long-duration failure. The following equation exhibits how to estimate the theoretical net throughput rate (TR_{net}) of the system with a buffer between the two subsystems:

$$TR_{net} = Min\{TR_1 \times A_1 \times B_1, TR_2 \times A_2 \times B_2\}.$$

Here B_1 and B_2 are the buffer impact factors, presenting how much the buffer reduces the corresponding blocked time and starved time. They can be estimated as:

$$B_1 = \frac{1 - \left[1 - \sum_0^{Max}\left(P_{WIP} \times \sum_0^{Max-WIP} P_{DT-2}\right)\right] \times (1 - A_2)}{A_2},$$

$$B_2 = \frac{1 - \left[1 - \sum_{Min}^{Max}\left(P_{WIP} \times \sum_{Min}^{WIP} P_{DT-1}\right)\right] \times (1 - A_1)}{A_1},$$

where, P_{WIP} is the probability of having WIP units in a buffer, P_{DT-1} and P_{DT-2} are the probability distributions of the downtime of Subsystems 1 and 2, respectively. Max is the conveyor capacity/buffer size, and Min is the required minimum buffer size (in terms of a number of WIP units), which are explained in Chapter 3. More discussion follows.

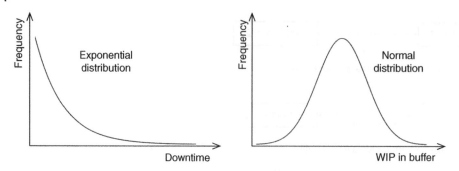

Figure 8.8 Probability distributions for system downtime and WIP in buffer.

8.2.1.2 Estimates Discussion

Following the previous discussions, the downtime of a system is assumed following an exponential distribution (Figures 6.15–6.17), and the WIP in a buffer is assumed following a normal distribution (Figure 3.17). Figure 8.8 depicts their respective distribution curves.

From the equations of B_1 and B_2 in the previous subsection 8.2.1.1, two observations emerge. First, if there is no buffer in the system, $P_{WIP} = 0$. Then both B_1 and B_2 are equal to one, meaning no buffer effect.

Second, the buffer has an infinite capacity (Max $= \infty$), $P_{WIP} = 1$. The maximum values of the impact factors are: $B_1 = \frac{1}{A_2}$ and $B_2 = \frac{1}{A_1}$, meaning the largest possible buffer effects. For instance, if $A_1 = 0.95$, the calculation yields $B_2 = \frac{1}{A_2} = \frac{1}{0.95} \approx 1.05$.

Using the subsystem parameters in Table 8.1, with Max $= 12$ and Min $= 2$, the calculation yields $B_1 = 1.067$ and $B_2 = 1.048$. Note, the calculation requires minor coding in Excel or other software based on the equations. The theoretical net throughput rate of the entire system with a buffer is approximately:

$$TR_{net} = \min\{59 \times 94.5\% \times 1.067, 60 \times 93.5\% \times 1.048\} \approx \min\{59.5, 58.8\} = 58.8 \text{ (JPH)}.$$

Compared with the system without the buffer (TR_{net} is 55.8 JPH), adding the buffer exhibits an improvement of 3.0 JPH or 5.1%. Although the improvement is an estimate, it provides a useful reference for considering buffer size and cost, etc.

It is important to note that such an estimation is based on statistics under the assumptions mentioned, without considering the variation, real-time interactions, and other actual factors. For instance, if downtime and WIP in the buffer follow different distributions, the calculation results of B_1 and B_2 would be different. Therefore, the estimate may be used for comparison purposes, but not for an exact prediction.

Subsection Summary: This subsection introduces a statistical approach to estimate the through-put improvement by adding a buffer between subsystems. It provides equations to estimate buffer impact factors with a numerical example under certain assumptions and limitations. The estimation can serve as a reference for buffer design.

8.2.2 Total Cost Analysis

8.2.2.1 Buffer Effect Diminishment

Buffers offer benefits to production throughput, although they also incur costs. How large they should be in a particular manufacturing system can be analyzed and justified based on financial analysis or throughput accounting, as discussed in Chapters 2 and 4.

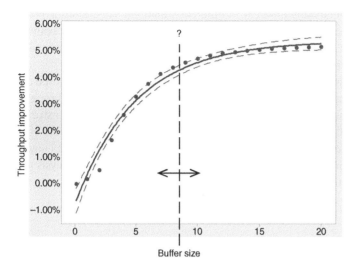

Figure 8.9 Exemplified relationship between buffer size and throughput improvement.

In most cases, the conveyor's transfer function is essential for the manufacturing system to work. The focus is on the conveyor's buffer effect, particularly the effect of buffer size, on throughput performance.

One primary purpose of a buffer is to cover small downtime events and variations. For instance, the goal of a buffer is to cover 80% of the downtime of a system. In such cases, determining buffer size is straightforward with the known or predicted downtime characteristics.

If buffers are used to improve throughput improvement, determining the optimal buffer size for throughput improvement requires closer analysis and discussion. An illustration of the impact of different conveyor (buffer) capacities on the throughput performance of a specific manufacturing line is presented in Figure 8.9, derived from computer simulation.

While the specific values on the axes of Figure 8.9 apply to this particular case, the general trend of improving throughput is about the same. The specific quantitative relationship between a buffer and throughput performance is previously discussed in subsection 8.2.1.

As buffer size increases, the buffer effect on throughput performance significantly diminishes beyond a certain size. The buffer size is linked to the upfront investment for the conveyor systems. Therefore, it is not necessary that the larger the buffer, the better the throughput is because of limited resources. The crucial question is how to determine the optimal buffer size, as shown by the vertical dashed line in Figure 8.9.

8.2.2.2 Cost and Benefit Analysis

Therefore, an appropriate buffer effectively supports system throughput while maintaining reasonable costs. The buffer benefits on throughput, increasing actual production time by reducing starved time and blocked time, can be quantified monetarily.

The costs of buffers include the initial investments and operating expenses of the conveyor systems and WIP carriers, the cost of WIP units, and operating expenses. Conveyor costs are influenced by its length, technology used, and complexity. The cost of WIP carriers is typically calculated per unit. Operating expenses encompass maintenance and spare parts.

Then, the total cost for justifying buffer utilization and sizing can be calculated as follows:

Total cost = Throughput benefits + Buffer costs.

A total cost curve for a buffer should have the lowest cost point in the long term, as depicted in Figure 8.10. The curve can be remarkably different in different situations. The declining portion before the lowest point shows increasing buffer size can reduce the total cost. After the point, the total cost climbs because of higher buffer costs with diminishing buffer effect. A total cost analysis helps determine the size of buffers.

Figure 8.11 shows another analysis, where a larger buffer size can have a shorter break-even time. In some cases, the conclusions can be the opposite, depending on conveyor costs and throughput financial benefits. Therefore, conducting a specific study to determine the optimal buffer size. The design team can then decide on the buffer size accordingly.

Subsection Summary: Optimizing buffer size entails assessing total costs, including conveyor costs and throughput financial gains, a relationship that can be complex. Total cost analysis helps facilitate a buffer size design for throughput excellence.

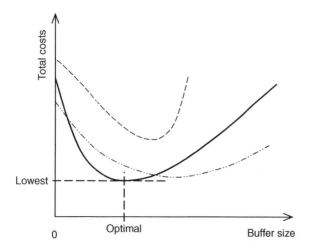

Figure 8.10   Total cost curves with buffer size (with optimal size for cost minimization).

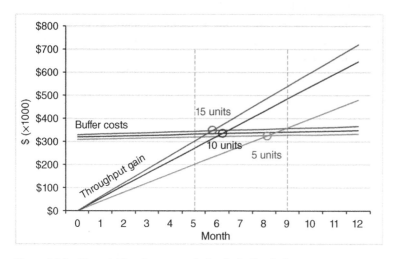

Figure 8.11   Financial break-even analysis of a buffer design.

8.2.3 Research Perspectives

8.2.3.1 Research Approaches

The optimal allocation of buffers and their sizes between workstations has been a subject of study for decades, known as the buffer allocation problem (BAP), initially defined by Koenigsberg [1959]. The BAP research continues to be an area of active research [Xi et al. 2020].

Weiss et al. [2019] conducted a comprehensive literature review covering 183 published studies on BAP. The literature review categorized BAP studies based on three main aspects:

- Process modeling assumptions: Includes factors such as the reliability of workstations, uniqueness of workstations, and line balance status. Recent studies placed emphasis on unreliable workstations, more complex lines with over ten workstations, and unbalanced lines.
- Modeling objective function and constraints: Involves considerations of throughput performance, CT, and the number of WIP. Some studies incorporated profit considerations, while others aimed at minimizing WIP and various performance measures.
- Study methods: Pertains to the methodologies employed such as exact numerical evaluations, integrated optimization, iterative optimization, and simulation-based evaluations.

As a major approach, iterative optimization methods construct a mathematical model that integrates a solution creation method with a system performance evaluation method [Demir et al. 2014]. Generative methods search for buffer allocations that meet constraints. This involves generating different solutions in each iteration. Evaluation methods assess system throughput performance based on analysis and simulation results. Both methods collaborate through a feedback loop (see Figure 8.12), continuing until the optimal solution is found or pre-determined criteria are met.

While research findings offer general guidelines, applying them to real-world systems with their complexities and unique features requires additional work. This underscores the need for further, customized buffer solutions, bridging the gap between theoretical optimization and practical system design.

8.2.3.2 Technical Challenges

Understanding the technical challenges of solving and optimizing BAP is essential for advancing research and bridging the gap between theory and practice. The challenges from real-world manufacturing systems include the following:

- The complexity of these problems grows exponentially with the system's size and the number of variables, making finding the optimal solution computationally intensive.
- There are some uncertainties as some influencing variables are unknown, requiring assumptions that may or may not accurately reflect real situations, potentially leading to less accurate BAP analysis outcomes.
- Evaluating and validating system performance during development presents a challenge due to the lack of readily available real systems. Only a handful of existing studies validated their findings with actual manufacturing systems.

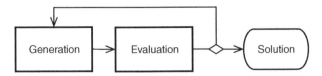

Figure 8.12 Integrated generation-evaluation process for buffer allocation.

- Several methods exist including aggregation approaches, decomposition methods, expansion, approximate methods, and Markovian modeling. Integrating them is changing and opportunities.
- Current modeling methodology may not meet the fast-paced demands of manufacturing system development. Addressing this time constraint is crucial for practical application.

These challenges contribute to a significant gap between research findings and practical implementation in manufacturing development. Many studies use a limited number of workstations and simplified assumptions, as seen in a recent study that only considered systems with up to five workstations [Boulas et al. 2021]. In contrast, real-world systems often comprise hundreds of diverse workstations.

Bridging this gap requires collaborative projects involving both experienced manufacturing practitioners and researchers. Such partnerships can lead to practical, validated solutions optimized for the needs of real-world manufacturing environments.

Subsection Summary: This subsection reviews the research approaches for BAP, emphasizing the need to acknowledge technical challenges and promote collaborative projects to bridge the theory–practice gap in manufacturing system design.

8.3 Throughput Capability Design

8.3.1 Cycle Time Design Considerations

8.3.1.1 Components of CT

Workstation CT design assigns specific manufacturing operations. As depicted in Figure 8.13, the design CT of a workstation includes the following four elements:

- Processing time: The time required to complete all operations.
- Time tolerance: Accounts for variations due to machines, materials, etc. An estimate of 3–5% of the processing time may be considered. If the variation is known from similar situations, the tolerance can be set to two standard deviations of the known variation.
- Time buffer: Specific for some workstations to mitigate potential small delays in the process. Its necessity depends on potential system issues and buffer analysis.
- Spare time: Unused time within the allotted design CT, either due to technical constraints (unwanted) or intentionally reserved for future use.

While time tolerance and time buffer serve different purposes, they can be combined in practice. They should be determined through analysis and/or computer simulation, not simply as a conservative measure. The values of time tolerance and time buffer should be justifiable based on the

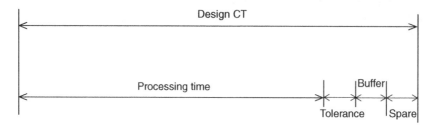

Figure 8.13 Time elements in workstation cycle time design.

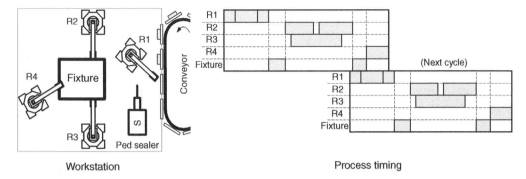

Workstation   Process timing

Figure 8.14 Process steps in parallel and overlapping in the workstation process.

unique requirements of each workstation. In the cases of sufficient spare time, both time tolerance and time buffer may become unnecessary.

The objective of workstation CT design is to ensure that the processing time is smaller than the design CT while appropriately considering the other three time elements. The focus of CT design efforts is typically directed toward reducing processing time.

Both product design and process planning play a crucial role in the selection of manufacturing processes, equipment, and workflows, all of which influence the processing time required. In process planning, the introduction of additional resources, adoption of new technologies, and workload adjustments can lead to a decreased processing time.

Additional opportunities exist to improve the workstation CT in process planning. For example, some process steps or tasks can run concurrently and/or overlap. Figure 8.14 shows an example of parallel tasks within a cycle and partial overlapping of consecutive cycles.

8.3.1.2 CT Design Adjustment

Process planning seeks balanced workloads across workstations while meeting their CT requirements. This ensures all workstations can complete their tasks within the allotted time simultaneously, enabling smooth production flow.

Figure 8.15 illustrates a simplified example of a serial system with eight, each having a design (required) CT of 45 seconds, with no time tolerance or buffer in this case for simplicity.

Table 8.2 presents the initial workload distribution, revealing an overload in Workstation 5 that requires adjustment. One approach is to shift a two-second workload from Workstation 5 to Workstation 6. Table 8.3 displays the revised CT values (shaded), aligning with the design CT.

In practical scenarios, achieving uniform CT for all workstations can be challenging, or even impractical, due to technical constraints, such as specific sequence requirements. In addition, the designed CT may require additional revisions to accommodate other factors, which will be discussed in the following subsections.

8.3.1.3 System Configuration and CT

As discussed previously, system configurations can be serial, parallel, and hybrid layouts. The configuration of workstations in a system significantly influences CT design and throughput analysis.

Figure 8.15 Example of a serial system with eight workstations.

Table 8.2 Workload design example of workstations (initial).

Workstation	1	2	3	4	5	6	7	8
Design CT (seconds)	45	45	45	45	45	45	45	45
Initial design CT (seconds)	40	41	39	44	46	38	43	39
Initial workload (%)	89	91	87	98	102	84	96	87

Table 8.3 Workload design example of workstations (revised).

Workstation	1	2	3	4	5	6	7	8
Revised design CT (seconds)	40	41	39	44	44	40	43	39
Revised workload (%)	89	91	87	98	98	89	96	87

For instance, a product with a total processing time of 650 seconds and a required throughput rate (TR) of 60 JPH. In a serial system configuration, the required workstation CT can be estimated as follows:

$$CT = \frac{3600}{TR} = \frac{3600}{60} = 60 \text{ (seconds)}.$$

The WIP transfer time between workstations is ten seconds, which is included in the workstation CT. Other supporting functions, like tool changing and material handling, operate in parallel with processing time and do not incur additional time.

If designed with a serial configuration, each workstation would have 50 out of the required 60-second CT for manufacturing processing. The system would require at least 13 workstations, calculated as:

$$\frac{650 \text{ seconds}}{50 \text{ seconds}} = 13 \text{ (workstations)}.$$

Figure 8.16 presents the CT and throughput time for this serial system design.

Alternatively, a parallel configuration could be used, for example, with two parallel segments labeled A and B. The analysis of cycle time and throughput time differs from that in the serial configuration, see Figure 8.17.

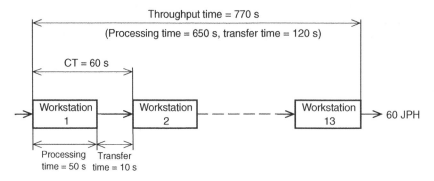

Figure 8.16 Cycle time and throughput time in a serial configuration.

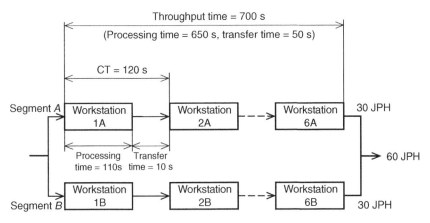

Figure 8.17 Cycle time and throughput time in a parallel configuration.

In this parallel design, each segment shares half the workload, resulting in a TR of 30 JPH for each segment. The workstation design CT is then:

$$CT = \frac{3600}{TR} = \frac{3600}{30} = 120 \text{ (seconds)}$$

Therefore, six workstations, with a total of 660 seconds of processing time (calculated as 6×110), are needed for each segment. This leaves ten seconds of extra processing time for each segment. More importantly, this parallel design needs only 12 workstations in total, one less than the serial configuration.

In general, a parallel layout requires fewer workstations or allows for additional processing time compared to a serial configuration, while meeting the same manufacturing throughput requirements. However, parallel configurations can be more complex and expensive to implement and challenging to manage quality discussed in subsection 4.3.2 of Chapter 4.

Hybrid configurations combine elements of both serial and parallel layouts. They can be analyzed using the same principles applied to serial and parallel configurations, considering the workload distribution occurs across the different segments.

8.3.1.4 Workstation CT Influence

The theoretical relationship between CT and gross TR ($CT = \frac{3600}{TR}$), which is discussed in Chapter 1, assumes a system with perfectly reliable and no variability. However, nonperfect reliability and variations lead to interactions among process steps and workstations such as starvation and blockage (refer to subsection 7.3.2 of Chapter 7). While these interactions are often small and unnoticeable, they render the theoretical analysis less accurate. Computer discrete event simulation (DES) offers a more accurate estimation method, considering real-time and random interactions.

In Chapter 7, the interaction results due to reliability are discussed. A similar discussion from CT can be conducted, focusing on the influence of different workstation CTs. For this, an eight-workstation serial system, with all workstations having a reliability of 98%, is considered. Seven workstations have a CT of 60 seconds, while one workstation has a CT of 63 seconds, 5% slower than others.

In this example, there is one slow workstation. In theory, it affects the system output in the same way and at the same level, regardless of its position in the system. To investigate, we placed the slow workstation at different positions in the system and ran computer DES eight times for each configuration, maintaining identical conditions. Figure 8.18 displays the simulation results.

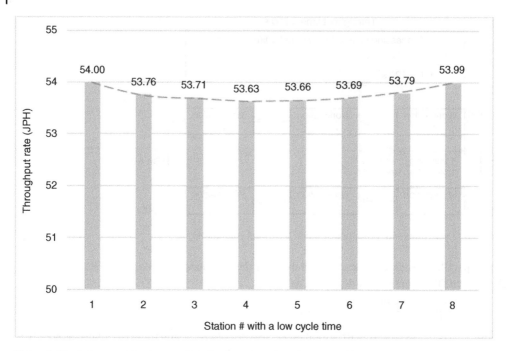

Figure 8.18 Influence of a slow workstation at various locations on throughput.

In this case, as the slow workstation's position varies in the system, the impact is not the same. Its influence is approximately 0.7% different when in the center compared to its placement at the beginning or end of the system. This finding is similar to the earlier discussion on workstation reliability effects in Chapter 7.

Subsection Summary: This subsection delves into key aspects of workstation CT design, including processing time, time tolerance, time buffer, system configurations, and internal interactions. It highlights the limitations of theoretical calculations and valuable computer simulations for precise impact assessments of these factors on system performance.

8.3.2 Throughput Capability Balance

8.3.2.1 Importance of Capability Balance

Achieving operational balance and minimizing inherent bottlenecks share the same fundamental goal. The aim of a balanced system design is to ensure that all functions and elements, including workstations, equipment, and human workers, have a similar workload in terms of time and throughput capacity for a smooth process flow. A balanced system can operate smoothly without significant blockage and starvation. Line balancing, or design for balance, aims to ensure even workload distribution among elements for smooth production flow and to prevent bottlenecks.

Each workstation in a system has a unique level of reliability, CT, and quality, collectively influencing operational performance. Table 8.4 is a continuation of Table 8.3 with additional calculated gross TR ($TR = \frac{3600}{CT}$) and the estimated/known A (operational availability).

A primary design requirement is net TR. If the system's operational availability (A) is the only concern, the relationship between TR_{net} and TR_{gross} (discussed in subsection 7.2.1 of Chapter 7) is:

$$TR_{net} = TR_{gross} \times A.$$

Table 8.4 Throughput capability of workstations based on cycle time design.

Workstation	1	2	3	4	5	6	7	8
Revised design CT (seconds)	40	41	39	44	44	40	43	39
Gross TR (JPH)	90.0	87.8	92.3	81.8	81.8	90.0	83.7	92.3
Estimated A (%)	97	93	94	91	88	94	93	98
Net TR (JPH)	87.3	81.7	86.8	74.5	72.0	84.6	77.9	90.5

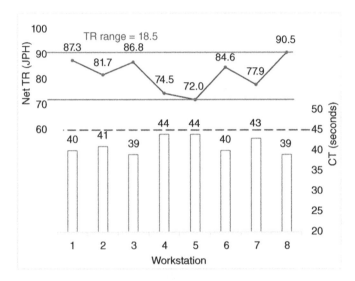

Figure 8.19 Example of system CT design results (revised).

Accordingly, the net TR of each workstation is determined, listed in the last row of Table 8.4 and illustrated in Figure 8.19.

At this point, the design satisfies the design CT requirement of 45 seconds. However, the net TR of the system is at approximately 72 JPH, as Workstation 5 is the lowest performer, acting as the bottleneck in production. Furthermore, the figure shows a significant variation in net TR, with a range of 18.5 JPH. Such an imbalanced capability would lead to hindering smooth system operations. This prompts the necessity of improving system capability balance and mitigating inherent bottlenecks. The discussion continues in the next subsection 8.3.2.2.

8.3.2.2 Balance of CT and Reliability

Both CT and reliability of workstations are pivotal attributes influencing throughput performance. Three predominant tactics for enhancing throughput capacity are as follows:

- Decreasing CT
- Increasing availability or reliability
- Applying both concurrently

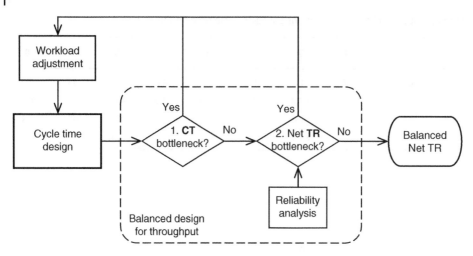

Figure 8.20 Design process flow for a balanced net throughput rate.

As previously discussed, reliability often translates to operational availability, and CT is inversely related to TR. Depending on specific situations, addressing either or both can effectively improve throughput.

In designing system capacity with a focus on its balance, start by addressing the CT or workload of each workstation, aligning with conventional design practices. Additional efforts are directed towards bottleneck avoidance to enhance throughput performance. Figure 8.20 illustrates the process with two focal points of this design approach in the following two steps:

1. Reviewing CT design and identifying the possible bottlenecks of a system in terms of CT, realigning workloads, if necessary and feasible, to achieve a CT balance (as seen in Table 8.3).
2. Analyzing net TR and its distribution (the variation is shown in Table 8.4), adjusting individual workstation CT or workload to achieve a uniform net TR across the system.

When addressing both CT and reliability bottlenecks, iterative adjustments may be necessary to meet the objective.

Adjusting workload for throughput balance relies on the assumptions of known reliability of workstations and adjustability of workloads. From this perspective, using more reliable equipment to enhance operational availability is a prudent consideration in design.

Another consideration for this approach is that the system has only minor reliability issues. If significant downtime risks exist, other approaches, such as incorporating backup and/or redundant functions, as discussed in Chapter 7, should be considered simultaneously.

The following subsections provide more detailed explanations focusing on net TR with examples.

8.3.2.3 Consideration of Availability and OEE

Using the same example to continue the discussion, the CT of each workstation is revised again for reduced variation of net TR. The revised CT and net TR are presented in Table 8.5 and Figure 8.21. The net TR variation is reduced to 3.4 JPH from 18.5 JPH, signaling improvement but still indicating room for further optimization. This revised design has improved balance with the theoretical system throughput of 79.7 JPH, much better than the first revision of 72 JPH.

Operational availability is often the major concern, and additional factors may need to be considered. In such cases, using OEE, including speed performance (P) and quality rate (Q), along with operational availability, is a viable choice, if P and Q information is available during design.

Table 8.5 Throughput capability of workstations based on revised cycle time design.

Workstation	1	2	3	4	5	6	7	8
Revised2 CT (seconds)	42	41	42	40	39	41	42	43
Gross TR (JPH)	85.7	87.8	85.7	90.0	92.3	87.8	85.7	83.7
Estimated A (%)	97	93	94	91	88	94	93	98
Net TR (JPH)	83.1	81.7	80.6	81.9	81.2	82.5	79.7	82.0

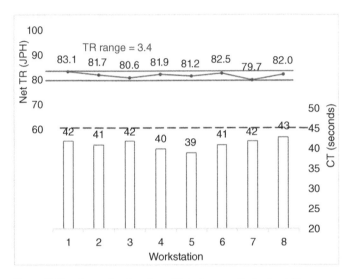

Figure 8.21 Example of system CT design results (revised 2).

For OEE-based net TR calculation, the formula is:

$$TR_{net} = TR_{gross} \times OEE.$$

Using OEE-based net TR and following the sample principle, an ideal design is that all work-stations have the same net TR as a bottleneck-free system. Using OEE can be more challenging than using A because more variables are involved, may not be fully understood or their data is not available during design phases.

It can be challenging or even impractical to have all workstations with the same net TR due to technical, time, and resource constraints. Therefore, a small variation range of TR, for example, 2 JPH, may be acceptable for some cases. A manufacturing system, designed to run for years, benefits from smooth operations, offsetting the design investments in a brief period.

8.3.2.4 Thoughts about Balanced Design

Balanced design can also be achieved through two methods requiring additional investment: capacity buffering and time buffering. Capacity buffering involves adding extra equipment or personnel to increase throughput capacity. This can be useful when addressing inherent process limitations like slowness or low quality. For example, by producing more units than needed, the system can offset the impact of defective units without hindering overall output.

Time buffering, on the contrary, involves adding extra time to the process cycle to account for potential delays or rework. For a process prone to producing defects, a time buffer allows for additional work time for quality checks and adjustments if needed.

Balanced design, also known as assembly line balancing (ALB), is a topic of interest for system design professionals and researchers [Eghtesadifard et al. 2020; Chutima 2020]. However, ALB presents a complex optimization problem in theory, involving different objectives, such as maximizing throughput, minimizing total cost, and lead time, among others. These optimizations must be achieved while considering real-world constraints like equipment capacity, processing time, and process flow.

One promising approach to address this complexity is multi-objective optimization. This method evaluates multiple solutions to simultaneously optimize each objective or stage under a set of constraints. Many research papers on this subject include studies on small-scale manufacturing systems, such as Gao and Zhang [2015], and large systems, such as engine manufacturing lines [Chica et al. 2019]. However, few general approaches translate directly to large, complex manufacturing environments. This presents an opportunity for future research.

Furthermore, current ALB solutions often struggle with the numerous variables and dynamic conditions encountered in real-world production. For example, Dörmer et al. [2015] found that most ALB models assume a fixed line balance even when producing mixed products, an unrealistic assumption for high-variant assembly lines. In addition, many ALB studies make unrealistic assumptions about perfect reliability, flawless quality, and ideal data distributions, which rarely reflect the realities of current production environments.

Subsection Summary: Designing a balanced manufacturing system requires addressing bottlenecks through optimizing cycle time and reliability and leveling throughput capacity across workstations. Iterative adjustments, considering constraints and buffers, facilitate a smooth workflow and improve throughput performance.

8.4 Additional Considerations

8.4.1 Value Creation and Throughput

8.4.1.1 Value Categories

In addition to the previously discussed key elements of CT, reliability, and balancing, the value creation for end customers is another consideration for system design. Understanding the value of each manufacturing task and process step is the first step toward reviewing and enhancing system productivity in value creation. The subsequent goal is to eliminate waste and minimize non-value-added activities that are necessary or unavoidable. This approach ensures that resources and time are efficiently utilized for value-added processes.

Value creation follows the lean manufacturing principles. All activities in a manufacturing system can be classified into the following three groups:

- Value added (VA): These manufacturing activities contribute to the product functionality and quality for end customers such as machining and assembly processes.
- Necessary non-value-added (NNA): These activities do not add value to end customers but are essential for some operations such as conveyor transferring WIP units, tools changing, and quality inspection.
- Waste (W): These activities consume resources but are not meaningful to customers and businesses such as waiting time (starvation or blockage), unused resources (equipment, workforce, etc.), excess inventory, and quality defects.

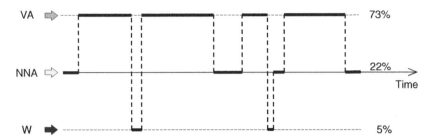

Figure 8.22   Value assessment chart for a manufacturing workstation.

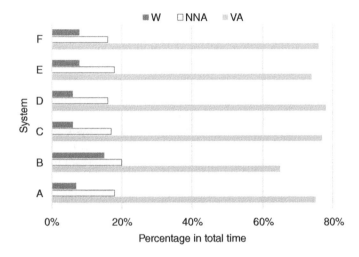

Figure 8.23   Value assessment summary of manufacturing systems.

Value evaluation can reveal the effectiveness of a manufacturing system. A value assessment chart can illustrate manufacturing activity values at different operational levels including workstations, lines, or systems. Figure 8.22 shows an example of a value assessment chart for a workstation. In this example, the VA operation accounts for 73% of the total production cycle time.

In addition, value assessment can be illustrated through charts at system levels. In the case of a large and complex system, a bar chart can summarize and compare the value assessment of subsystems, as illustrated in Figure 8.23. In the figure, distinct colors represent the following value categories: green for VA, yellow for NNA, and red for W activities. The assessment offers guidance to minimize NNA activities and eliminate W activities in system design.

8.4.1.2   Discussion on Value Evaluation

After waste activities are identified, most of them can be eliminated, or at least reduced. However, NNA activities may require more careful reviews to see if they can be reduced or eliminated. If these activities are deemed essential, they should operate concurrently with value-added activities. This approach optimizes NNA activities, enhancing the effectiveness of the manufacturing system.

As a type of waste, waiting time, being blocked or starved by neighboring systems and/or external factors, can be challenging to address. Consistent or significant waiting indicates an unbalanced system design, as discussed earlier, and its resolution may or may not be achievable through production control.

A challenge in the value assessment is to identify NNA activities. Different departments may have varying perspectives. These activities can be important for several reasons. For instance, while quality inspections may not improve product quality or functionality, they are crucial for assessing and ensuring product quality, thus contributing to customer satisfaction. At times, it plays a role in ensuring the smooth progression of downstream manufacturing operations.

Associated with NNA activity identification, there is a debate on whether buffers and their WIP units should be considered non-value added for end customers. WIP buffers, in addition to serving essential transfer functions, can reduce the variability of system operations, improving throughput performance. Therefore, considering buffers as "non-value added" oversimplifies the perspective. Thus, it contributes value to the business, even though it may not directly impact end customers. Many lean manufacturing experts concur that a theoretical just-in-time (JIT) scenario with zero inventory can be disastrous if something goes wrong, potentially causing production systems to shut down [Cusumano et al. 2021].

Numerous studies assess the value of manufacturing systems. As an illustration, a case study demonstrated that a revised plant layout reduced the total time from order placement to shipment by 75% in a rope manufacturing system [Yuvamitra et al. 2017]. Similarly, another study proposed a method to address customer needs in complex and successive engineering phases [Mohr et al. 2020]. Conversely, another study simultaneously considered value creation, throughput (OEE), economic, and environmental performance [Castiglione et al. 2022]. Another analysis assessed resource consumption based on its contribution to value generation [Papetti et al. 2019]. Another article introduced life cycle assessment and DES for performance analysis [Samant and Prakash 2021].

Subsection Summary: The subsection explores value assessment in manufacturing system design, highlighting the importance of identifying VA, necessary NVA, and W activities. It discusses challenges, such as waiting time and buffer value, and displays studies for system optimization.

8.4.2 Additional Key Aspects

8.4.2.1 Design for Quality Repair

Quality assurance is critical not only for product quality but also for both throughput excellence and value creation. As discussed in previous chapters, addressing quality during the product design phase and manufacturing process planning phase can help prevent or minimize most quality issues and defects in production.

Quality assurance can take various forms. As a fundamental one, quality inspections are used for a monitoring and measuring quality status in manufacturing systems. These inspections are placed at critical locations within the system to ensure that quality standards are being met. For critical quality attributes with potential defects, it is recommended to perform quality inspections immediately after the corresponding process; Figure 4.19 of Chapter 4 shows an example. This inspection approach helps detect and address quality issues earlier in the production process, rather than at the end of the production line where repairing defective units may be difficult and/or costly.

In conjunction with quality inspections, repair capabilities should be incorporated into systems by design. A rapid repair process can fix minor, easily correctable defects immediately after the inspection, with no or minimal disruption to the production flow. Figure 8.24 illustrates two scenarios of rejection repair after a quality inspection. The scenario in Figure 8.24a illustrates the repaired units are back to the line to have a re-inspection while the scenario in Figure 8.24b does

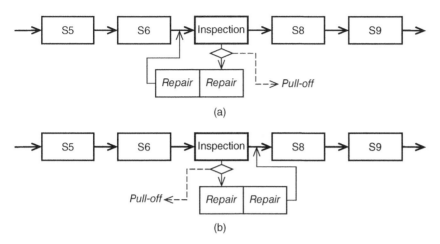

Figure 8.24 On-site repair functions integrated into a production line.

not require one. If defects are severe and repair takes time, the defective units need to be pulled off from the production line and sent to a dedicated repair area.

Figure 8.25 shows an example of repair functions, shaded in the figure, in a large manufacturing system of a vehicle paint shop [Tang 2018]. Certain processes, such as Underbody and Upper Seal in this case, are critical for quality attributes (water leak prevention and wind noise reduction for vehicles in this case). Specific quality assurance, namely, the repair processes (repair area and off-line seal in this example) are integrated into the system to ensure good quality WIP units proceed to subsequent processes.

Following repairs, the WIP units need to be reintroduced into the production lines, which can be challenging if products are being manufactured in a certain sequence, based on feature or color. Reinserting may require waiting to reach a specific point in a WIP flow, which needs to pause the system, reducing system throughput. It is crucial to design system functions for the reinsertion into the production process to maintain system throughput in high-volume production.

8.4.2.2 System Flexibility

The discussion so far focuses on a single product family with similar main manufacturing processes. However, businesses often face diverse and dynamic market demands. It is crucial to design a manufacturing system that is flexible to accommodate future new products with distinctive designs and processes. In this regard, system design enables one to adapt to changes in demand and stay competitive in the market.

The flexibility of a manufacturing system can be viewed and measured in several ways, including its ability to respond to internal and external changes. According to Grando et al. [2021], a manufacturing system's flexibility can be assessed through the following four key aspects:

- Mix flexibility: This refers to the ability to produce a wide range of products.
- Product flexibility or convertibility: The ability to convert existing machines and processes quickly and easily.
- Volume flexibility or elasticity: This refers to a system's ability to adapt to market variations while maintaining operational efficiency.
- Plan flexibility: About a system's ability to change production sequences.

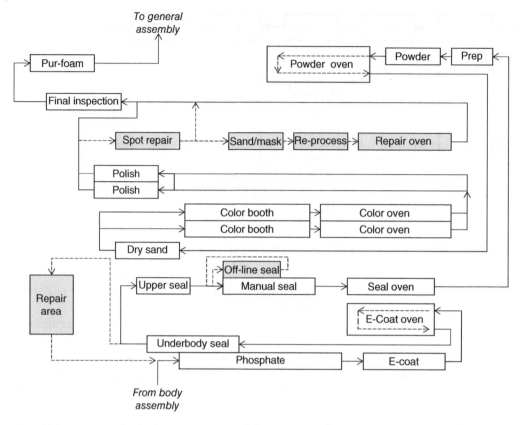

Figure 8.25 Integrated repair functions in vehicle paint systems (Tang, 2017/with permission of SAE International).

Therefore, retrofitting an existing manufacturing system is not always economically viable. Directly integrating flexibility into a new system is often more cost-effective.

The flexibility of a manufacturing system is highly dependent on the product design such as modular design and carryover of existing parts. There are numerous principles and research studies of design for manufacturing [Roxas et al. 2023]. Involving manufacturing professionals in product design is a way to effectively enhance product design for manufacturing.

In addition to product modular design, a manufacturing system should also be designed for modularity. Modular manufacturing is a research topic [Shaik et al. 2015; Brunoe et al. 2021] to make manufacturing systems more efficient, cost-effective, and resilient in the face of supplier changes.

Flexible manufacturing system design requires additional upfront investments, but it can bring long-term benefits in terms of operational cost-efficiency and adaptability to market changes. Figure 8.26 depicts an illustrative cost comparison of low and high-flexible vehicle assembly systems [Tang 2018]. High flexibility manufacturing systems can be modified without reinvesting in totally new systems, to adopt new products, leading to significant long-term savings.

The actual comparison can be analyzed based on the specific situation and data. However, justifying the initial investment and accurately predicting future market trends pose challenges, as inaccurate predictions could lead to suboptimal cost efficiency. Therefore, investing in flexibility brings a risk to achieving the best cost efficiency in system design and construction.

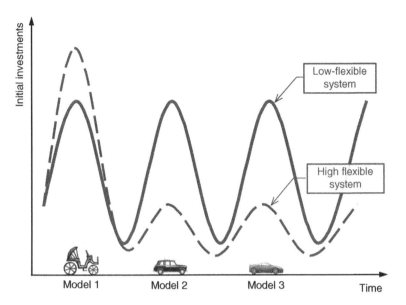

Figure 8.26 Cost implication of manufacturing system flexibility (Tang, 2017/with permission of SAE International).

8.4.2.3 Batch Process

More flexible than repetitive processes, a batch process is common in mid-variety product production. Batch processes can be designed into repetitive systems to make them more flexible. That means that products with similar characteristics are grouped to be processed, rather than randomly mixed and processed in a process flow. Within a batch, there may be limited variations such as using different process parameters, fixtures, and/or time. Figure 8.27a depicts a manufacturing system with batch processing. Figure 8.27b outlines the processes.

This example illustrates some important aspects of system design for batch processing, including the following:

- Designing a common process flow for the products going through different workstations.
- Addressing different CT and throughput times of each product batch.

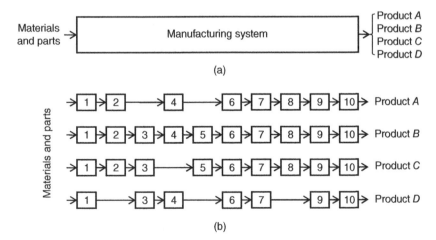

Figure 8.27 Batch process of a manufacturing system.

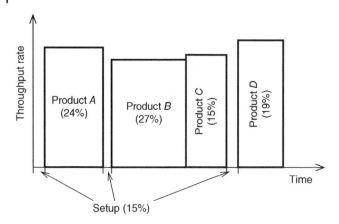

Figure 8.28 Batch throughput rates of a batch process system.

- Determining the system TR is based on individual batch TRs and production volumes. Figure 8.28 depicts an example. In this example, the overall system TR is:
  $$TR = (0.24 \times TR_A + 0.27 \times TR_B + 0.15 \times TR_C + 0.19 \times TR_D).$$
- Attempting to reduce the setup (nonoperation) time needed for product batches (15% in this example) to improve system throughput performance.

When selecting the batch size, there is a trade-off between throughput time, TR, equipment unitization, and labor productivity. This decision also considers factors such as production volume, product mix, production costs, and profit. For the technically feasible batching options, conducting a cost-benefit analysis can assist in comparing and deciding the batch size.

Figure 8.29 illustrates an example of how batching works in a large manufacturing system of a vehicle paint shop [Tang 2017]. Vehicles with assorted colors are in random customer orders in a production system. Color changes in a painting booth require purging paint with solvent and cleaning the old paint from the spray systems. Therefore, batching vehicles by color for painting processes has significant financial savings, environmental benefits, and regulatory compliance of

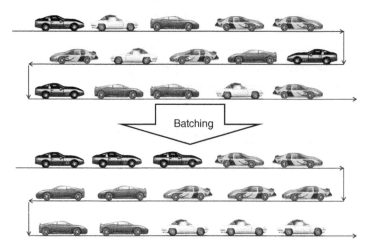

Figure 8.29 Batching process for vehicle paint operations (Tang, 2017/with permission of SAE International).

the paint shop operations. A common way to enable a batch process is to use a system conveyor. For example, in a paint shop, a conveyor system, called a sequence bank, sorts products according to color.

For products with distinct components, such as aerospace assemblies, system flexibility can be enhanced by making production lines reconfigurable [Jefferson et al. 2016]. A reconfigurable manufacturing system allows the process, machinery, line structure, and even layout to be rapidly adapted for a variety of products. This enables efficient production of highly varied products.

Subsection Summary: Effective manufacturing system design demands a strategic approach, addressing various aspects including quality assurance, flexibility, and batch processing. Achieving a balance between upfront investments and long-term benefits is crucial for sustaining manufacturing excellence.

8.4.3 Discrete Event Simulation

8.4.3.1 DES Overview

Manufacturing systems have a hierarchical structure that consists of systems, subsystems, sub-subsystems, etc. For example, a vehicle body shop may have 1000 robots and 20 assembly lines. Overall system performance depends on the real-time execution and interactions of all elements. These interactions are so complex and dynamic that they create a challenge for mathematical models to accurately capture their dynamic behaviors and predict a system's performance.

Computer discrete-event simulation (DES), a modeling technology and tool, can effectively handle the complexity and real-time interactions in manufacturing systems. DES describes the behavior of a system as a discrete series of events in time. Each event happens at a specific time and causes a change in the system state, which is the set of all element attributes. A process in a DES is a series of events that may involve activities. With DES, manufacturing systems can be modeled, analyzed, and optimized to enhance their performance.

DES has been around since the 1950s, continuously evolving and widely used in various industries. There are various applications of DES such as operation checks, throughput prediction [Al-zqebah et al. 2022], bottleneck prediction [Rocha and Lopes 2022], and layout optimization [Cisneros and Escobar 2022]. According to a comprehensive review, DES in the manufacturing industry has experienced rapid advances recently and is expected to remain a fruitful area for research and applications [Mourtzis 2020].

Furthermore, DES technology is not only an effective tool for new system development but also for supporting daily operational management. DES can be integrated into an operational management system to predict future operational effectiveness. An issue challenging DES analysis is the availability of production data, including information on minor stoppages and domain-specific DES data. A study addressed 11 dimensions of data quality for DES [Bokrantz et al. 2018]. These dimensions encompass accuracy, reputation, accessibility, currency, completeness, precision, relevance, resolution, traceability, clarity, and consistency. Current practices show that practical guidelines should be developed to support manufacturing companies in improving data quality in DES.

8.4.3.2 Throughput Analysis Example

DES is an effective tool for manufacturing design for throughput. For example, a DES with Matlab was conducted for the Philips system for a new shaving product [Strijker 2015]. The simulation suggested specific numbers of carriers (buffer sizes) inside the system and the size of the external buffer, where the locations of buffers are fixed based on the process flow. Under the recommended

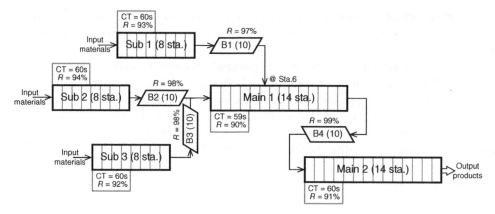

Figure 8.30 Example of a manufacturing system configuration.

buffer sizes, the simulation predicted the system throughput. Another study used simulation-based optimization to detect bottlenecks based on the active periods of an operation [Pehrsson et al. 2016].

For discussion and further exercise, Figure 8.30 demonstrates an example of a manufacturing system. This system has five subsystems (Sub 1, Sub 2, Sub 3, Main 1, and Main 2) connected with conveyors (B1–B4). The subsystems have multiple workstations, and each conveyor can hold up to ten WIP units. CT and reliability (R) are estimated based on system requirements and initial design. Figure 8.31 shows the corresponding model using DES software Simul 8.

Readers can use a DES software tool, such as Plant Simulation, Simio, or FlexSim, to build and run the model. By running the DES model, readers can change workstation parameters, like CT, reliability, buffer size and location, data distribution, and observe how they influence the system throughput performance. This simulation exercise remarkably enhances the understanding of system throughput and provides insightful guidance for optimizing system design for through put excellence.

DES has limitations. One limitation involves accuracy. The assumptions for DES modeling, for instance, reliability distribution, may not exactly match real-world situations. Consequently, DES results may not be the same as those from real production. Therefore, simulation engineer knowledge and assumptions used in modeling play a key role in DES accuracy. Another limitation is that DES is case dependent, not automatically comparing alternatives to find the optimal solution. This suggests the opportunities of integrating DES with optimization methodology, artificial intelligence, and/or other technologies. In addition, DES often requires a professional software package and trained engineers, which can be a challenge for small companies.

Subsection Summary: This subsection demonstrates DES as a powerful tool for manufacturing system analysis and design for throughput excellence. While DES presents certain challenges, including limitations in accuracy and reliance on professional software, it also has opportunities for integration with other technologies.

8.4.4 Advancing Manufacturing with Research

This book has uncovered a wealth of principles and methods, offering a roadmap for those navigating the intricate field of manufacturing excellence. While it does not provide all the direct answers, it sparks the journey toward mastery and creative application in real-world manufacturing scenarios. Moreover, these challenges serve as catalysts for ongoing research in this domain, aiming for a deeper understanding and the development of more effective tools.

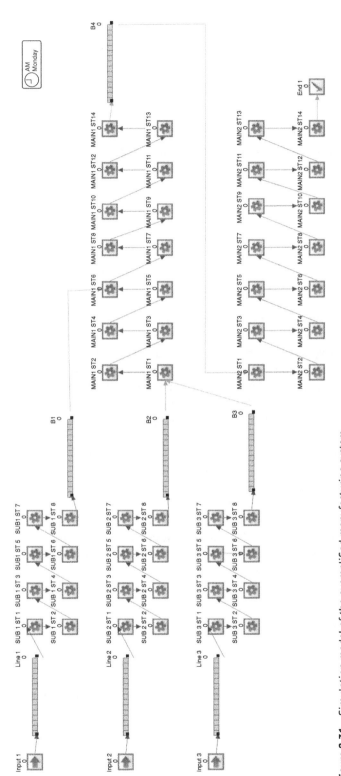

Figure 8.31 Simulation model of the exemplified manufacturing system.

System design remains a vibrant field of research, with numerous scholarly papers dedicated to optimizing manufacturing systems through innovative design approaches. This exploration often hinges on two key methodologies: analytical methods and simulation methods, which are occasionally fused together through integration for enhanced results.

Analytical methods, grounded in formal mathematical modeling and algorithmic solutions, provide a structured foundation for system design. For instance, certain techniques are built upon the Markovian process model [Papadopoulos et al. 2019], ontology [Arista et al. 2023], and recurrent neural networks [Chen et al. 2023]. Several studies investigated the buffer size. A study was conducted at a large food processing company that used a continuous process [Hossain and Sarker 2019]. The researchers developed and implemented a nonlinear integer program to optimize the tank (buffer) size, considering its holding, quality, process overshoot, and undershoot costs [Rajaram and Tian 2009].

Computer simulation stands as a robust tool capable of modeling intricate systems and capturing dynamic interactions inherent within them. Furthermore, it adeptly accounts for variables like reliability and CT, while offering the capacity to replicate real-world manufacturing operations. The increasing integration of 3D visual animations has augmented the utility of simulation in practical industrial applications. An exciting trend is the convergence of simulation and analytical techniques.

For instance, a recent study demonstrated the optimization of profits through a simulation-based metaheuristic approach tailored for high-mix, low-volume manufacturing [Herps et al. 2022]. In addition, novel technologies such as artificial intelligence (AI) [Arinez et al. 2020] have found their place in the realm of manufacturing system study and execution. However, the specifics of these technologies and their implementations are beyond the scope of this book.

Researchers continually seek to study manufacturing systems, acquiring a better understanding of their working situations and optimizing them through new mathematical modeling. For engineers and managers engaged in system design and throughput improvement on the production floor, it is essential not to get lost in theoretical details. Researchers should acquire hands-on experience with and understanding of specific processes and constraints, seeking better system designs and practical improvements. Experienced practitioners can contribute to future research by being closer to real-world manufacturing systems and constraints.

In conclusion, manufacturing excellence is an ongoing journey with its inherent challenges, and there is always a limitless horizon for innovation. Advancing manufacturing excellence involves creative applications of existing knowledge and the exploration of analytical models and computer simulations, driving continuous progress in the field.

Chapter Summary

8.1. Buffer Planning for Throughput

1. Buffers, such as WIP, capacity, and time buffers, absorb variabilities in manufacturing systems, ensuring smooth operations and optimal throughput.
2. Analyzing buffer size and total costs, including conveyors and throughput gains, is crucial for optimal system performance.

3. Strategic buffer allocation, for decoupling long lines, absorbing variation, and compensating for unreliable operations, is effective to enhance system throughput.
4. Large conveyor systems, acting as buffers, play a dual role in transporting WIP units and enhancing system efficiency.
5. The buffer effect on throughput performance is complicated, depending on several factors such as the number, location, and size of buffers.

8.2. Analysis in Buffer Design

6. Buffer effect can be estimated based on the assumptions of the exponential distribution of system downtime and the normal (or Poisson) distribution of WIP in a buffer.
7. Buffers provide limited throughput gains, introducing buffer impact factors (less than 10%).
8. Determine optimal buffer size by modeling total costs, combining conveyor, carrier, and operating expenses with throughput gains.
9. A break-even point for a buffer size can be estimated based on buffer investment and corresponding throughput gained.
10. Solving and optimizing BAP faces challenges like computational intensity, unknown variables, and the gap between research and practical application.

8.3. Throughput Capability Design

11. Workstation CT consists of four elements: processing time, tolerance, time buffer, and spare time.
12. Influence of product design and process planning on CT, with opportunities to optimize tasks and parallelize processes.
13. System configurations (serial, parallel, hybrid) affect workstation CT and system throughput time, with implications for the number of workstations needed.
14. Real-time interactions and variations impact system throughput, highlighting the need for computer simulation to estimate realistic performance.
15. Strategies for achieving balance capability in throughput by addressing CT, reliability, and minimizing bottlenecks for smooth system performance.

8.4. Additional Considerations

16. System design should address value by categorizing activities into value-added, necessary non-value-added, and waste, per lean principles.
17. Quality assurance, inspections, and repair capabilities are crucial for preventing defects, ensuring product quality, and maintaining throughput excellence.
18. Designing flexibility into manufacturing systems, measured through mix, product, volume, and plan flexibility, enables adaptation to dynamic market demands.
19. Batch process design requires considerations of common process flows, throughput times, and setup time reduction.
20. Computer discrete event simulation (DES) serves as a powerful tool for modeling, analyzing, and optimizing manufacturing systems, despite limitations related to accuracy and optimal solution finding.

Exercise and Review

Exercises

8.1 Two lines, linked in series with no buffer between them, have individual throughput rates of 55 and 56 JPH and operational availabilities of 92% and 90%, respectively. Estimate the throughput capacity at the end of the second line.

8.2 Two systems have TRs of 55 and 56 JPH and operational availabilities of 92% and 90%, respectively. A buffer between them has impact factors $B_1 = 1.037$ and $B_2 = 1.045$. Estimate the throughput capacity at the end of the second system.

8.3 A manufacturing line has seven workstations with a design CT of 50 seconds. The design CT for each workstation is 45, 42, 47, 50, 38, 40, and 43 seconds, respectively. Calculate the range of workloads for all workstations.

8.4 A manufacturing system has six workstations with a required net TR of 60 JPH. The design results are the gross TR and availability for each operation, listed in Table 8.6. Check the projected net TR for each workstation and determine if the system design meets its throughput requirement. What is the net TR range across all operations?

8.5 The individual availability (reliability) values of workstations in a production line are listed in Table 8.7. If the net TR is required to be at least 80 JPH, what is the maximum permissible CT for each workstation? (Hint: calculate required gross TR, then CT)

8.6 A manufacturing system has six workstations with different OEE capabilities, listed in Table 8.8. The design objective for the system is a net TR of at least 60 JPH. To achieve a good balance for the throughput performance, calculate the design CT for each operation.

Table 8.6 For Exercise 8.4.

Workstation	1	2	3	4	5	6
Design gross TR	61	65	64	65	66	62
Availability (%)	99	92	95	93	96	98 (%)

Table 8.7 For Exercise 8.5.

Workstation	1	2	3	4	5	6	7	8
Reliability (%)	94	98	97	94	96	95	99	94

Table 8.8 For Exercise 8.6.

Workstation	1	2	3	4	5	6
OEE (%)	93	88%	90	89	92	91

Review Questions

(The chapter covers these topics. For further discussion, it is recommended to seek additional information and examples. Diverse perspectives are encouraged.)

8.1 Describe and explain the general relationship between buffer size and throughput performance.

8.2 Discuss how to allocate a buffer to decouple a long system, aiming to achieve similar reliability across all segments.

8.3 Provide an example and discuss the application of the "variability buffering law."

8.4 Justify buffer allocation before and after a workstation with low reliability.

8.5 Discuss a situation where a conveyor shortcut path can boost throughput performance.

8.6 Comment on the assertion that the benefit of buffers on throughput performance is estimated to be less than 10% and explain why.

8.7 Explain how to decide a buffer size based on total cost, including conveyor costs and throughput gains.

8.8 Provide an example and examine a break-even assessment for buffer design.

8.9 Evaluate a research paper on buffer size or location, critique its approach, and examine its possible applications in manufacturing design.

8.10 Explain why a design with only equal CT (workload) is inadequate for system throughput, providing an example.

8.11 Discuss the necessity of time tolerance and time buffer for a specific application during cycle time design.

8.12 Discuss the interactions between workstations and their influence on system CT.

8.13 Explain two key factors of CT and reliability in a balanced design for throughput capability.

8.14 Discuss throughput balancing using OEE in system design, providing an example.

8.15 Review the evaluation of value creation (or addition) in system design, providing an example.

8.16 Provide an example of a manufacturing activity deemed as a necessary non-value-added task and discuss how to minimize it in system design.

8.17 Compare the characteristics of online and onsite-offline quality repair functions of a manufacturing system, providing examples.

8.18 Discuss the necessity and challenges of enabling manufacturing flexibility in design.

8.19 Provide an example and discuss a throughput improvement enabler for batch production.

8.20 Review a research paper on utilizing discrete event simulation in system design.

References

Al-zqebah, R., Hoffmann, F., Bennett, N., Deuse, J., and Clemon, L. 2022. Woolshed throughput improvement using discrete event simulation. Journal of Industrial Engineering and Management, 15(2), pp. 296–308.

Arinez, J.F., Chang, Q., Gao, R.X., Xu, C. and Zhang, J. 2020. Artificial intelligence in advanced manufacturing: Current status and future outlook. Journal of Manufacturing Science and Engineering, 142(11) p. 110804.

Arista, R., Zheng, X., Lu, J. and Mas, F. 2023. An Ontology-based Engineering system to support aircraft manufacturing system design. Journal of Manufacturing Systems, 68, pp. 270–288.

Bokrantz, J., Skoogh, A., Lämkull, D., Hanna, A. and Perera, T. 2018. Data quality problems in discrete event simulation of manufacturing operations. Simulation, 94(11), pp. 1009–1025.

Boulas, K.S., Dounias, G.D. and Papadopoulos, C.T. 2021. A hybrid evolutionary algorithm approach for estimating the throughput of short reliable approximately balanced production lines. Journal of Intelligent Manufacturing, 34(2), pp. 823–852. https://doi.org/10.1007/s10845-021-01828-6

Brunoe, T.D., Soerensen, D.G. and Nielsen, K., 2021. Modular design method for reconfigurable manufacturing systems. Procedia CIRP, 104, pp. 1275–1279.

Castiglione, C., Pastore, E., and Alfieri, A. 2022. Technical, economic, and environmental performance assessment of manufacturing systems: the multi-layer enterprise input-output formalization method. Production Planning and Control, 35 (2), pp. 1–18.

Chen, M., Furness, R., Gupta, R., Puchala, S. and Guo, W. 2023. Hierarchical RNN-based framework for throughput prediction in automotive production systems. International Journal of Production Research, 62(5), pp. 1–16.

Chica, M., Bautista, J., and de Armas, J. 2019. Benefits of robust multiobjective optimization for flexible automotive assembly line balancing. Flexible Services and Manufacturing Journal, 31(1), pp. 75–103.

Chutima, P., 2020. Research trends and outlooks in assembly line balancing problems. Engineering Journal, 24(5), pp. 93–134.

Cisneros, D., and Escobar, L. 2022. Plant Layout Selection Procedure Based on Discrete Event Simulation Software. In XV Multidisciplinary International Congress on Science and Technology, Springer, Cham, pp. 197–210.

Cusumano, M.A., Holweg, M., Howell, J., Netland, T., Shah, R., Shook, J., Ward, P. and Womack, J., 2021. Commentaries on "The Lenses of Lean." Journal of Operations Management, 67(5), pp. 627–639.

Demir, L. Tunali, S. and Eliiyi, D.T. 2014. The state of the art on buffer allocation problem: A comprehensive survey, Journal of Intelligent Manufacturing, 25(3), pp. 371–392. https://doi.org/10.1007/s10845-012-0687-9.

Dörmer, J., Günther, H.O., and Gujjula, R. 2015. Master production scheduling and sequencing at mixed-model assembly lines in the automotive industry. Flexible Services and Manufacturing Journal, 27(1), pp. 1–29.

Eghtesadifard, M., Khalifeh, M. and Khorram, M. 2020. A systematic review of research themes and hot topics in assembly line balancing through the web of science within 1990–2017. Computers & Industrial Engineering, 139, p. 106182.

Gao, Z. and Zhang, D. 2015. Performance analysis, mapping, and multiobjective optimization of a hybrid robotic machine tool. IEEE Transactions on Industrial Electronics, 62(1), pp. 423–433.

Grando, A., Belvedere, V., Secchi, R. and Stabilini, G. 2021. Production, Operations and Supply Chain Management, ISBN: 9788899902575, EGEA Spa — Bocconi UNIVERSITY Press.

Guide Jr, V.D.R. and Srivastava, R. 2000. A review of techniques for buffering against uncertainty with MRP systems. Production Planning & Control, 11(3), pp. 223–233.

Herps, K., Dang, Q.V., Martagan, T. and Adan, I., 2022. A simulation-based approach to design an automated high-mix low-volume manufacturing system. Journal of Manufacturing Systems, 64, pp. 1–18.

Hopp, W.J. and Spearman, M.J. 1995. Factory Physics: Foundations of Manufacturing Management, 1 ed., ISBN: 978-0256154641, McGraw-Hill Education, New York, NY.

Hopp, W.J. and Spearman, M.S. 2021. The lenses of lean: Visioning the science and practice of efficiency. Journal of Operations Management, 67(5), pp. 610–626.

Hossain, Md S. and Sarker, B. 2019. Optimum Buffer Size of Steel Strip Coils in a Continuous Pipe Production System. Transforming Decision Sciences Thought Emergent Technologies, 50th Annual Conference, New Orleans, LA, Nov. 23–25, 2019.

Imseitif, J. and Tang, H. 2019a. Effects Analysis of Internal Buffers in Serial Manufacturing Systems for Optimal Throughput, In Proceedings of the ASME 2019 14th International Manufacturing Science and Engineering Conference, ISBN: 978-0-7918-5874-5, MSEC2019-2912, 7 pages, June 10–14, 2019, Erie, PA, USA.

Imseitif, J., Tang, H., and Smith, M.G. 2019b. Throughput analysis of manufacturing systems with buffers considering reliability and cycle time using DES and DOE. Procedia Manufacturing, 39, pp. 814–823. (ISSN: 2351-9789)

Jefferson, T.G., Benardos, P. and Ratchev, S., 2016. Reconfigurable assembly system design methodology: a wing assembly case study. SAE International Journal of Materials and Manufacturing, 9(1), pp. 31–48.

Koenigsberg, E. 1959. Production lines and internal storage—A review. Management Science, 5(4), pp. 410–433.

Mohr, F., Rübel, P. and Ruskowski, M. 2020. A holistic approach for value-added interaction modeling in flexible manufacturing systems. Procedia Manufacturing, 51, pp. 1245–1250.

Mourtzis, D. 2020. Simulation in the design and operation of manufacturing systems: state of the art and new trends. International Journal of Production Research, 58(7), pp. 1927–1949.

Papadopoulos, C.T., Li, J. and O'Kelly, M.E.J. 2019. A classification and review of timed Markov models of manufacturing systems. Computers and Industrial Engineering, 128, pp. 219–244. https://doi.org/10.1016/j.cie.2018.12.019.

Papetti, A., Menghi, R., Di Domizio, G., Germani, M. and Marconi, M. 2019. Resources value mapping: a method to assess the resource efficiency of manufacturing systems. Applied Energy, 249, pp. 326–342.

Pehrsson, L., Ng, A. H. and Bernedixen, J. 2016. Automatic identification of constraints and improvement actions in production systems using multi-objective optimization and post-optimality analysis. Journal of Manufacturing Systems, 39, pp. 24–37.

Rajaram, K. and Tian Z. 2009. Buffer location and sizing to optimize cost and quality in semi-continuous manufacturing processes: Methodology and application, IIE Transactions, 41 (12), pp. 1035–1048. https://doi.org/10.1080/07408170902889694.

Rocha, E.M. and Lopes, M.J. 2022. Bottleneck prediction and data-driven discrete-event simulation for a balanced manufacturing line. Procedia Computer Science, 200, pp. 1145–1154.

Roxas, C.L.C., Bautista, C.R., Dela Cruz, O.G., Dela Cruz, R.L.C., De Pedro, J.P.Q., Dungca, J.R., Lejano, B.A. and Ongpeng, J.M.C. 2023. Design for manufacturing and assembly (DfMA) and design for deconstruction (DfD) in the construction industry: Challenges, trends and developments, Buildings, 13(5), p. 1164.

Samant, S. and Prakash, R. 2021. Innovative framework for lean and green complex manufacturing systems using value stream mapping. International Journal of Advanced Operations Management, 13(3), pp. 292–311.

Shaik, A.M., Rao, V.K. and Rao, C.S. 2015. Development of modular manufacturing systems—A review. The International Journal of Advanced Manufacturing Technology, 76, pp. 789–802.

Stratton, R. 2008. Theory building: relating variation, uncertainty, buffering mechanisms, and trade-offs. In The World Conference on Production and Operations Management, Tokyo, Japan, August.

Strijker, V. 2015. Using simulation for a new production line at Philips Drachten Master's Thesis, University of Twente.

Tang, H. 2017. Automotive Vehicle Assembly Processes and Operations Management, ISBN: 978-0-7680-8338-5, SAE International, Warrendale, PA.

Tang, H. 2018. Manufacturing System and Process Development for Vehicle Assembly, ISBN: 978-0-7680-8346-0, SAE International, Warrendale, PA.

Turner, A. 2020. Evaluation of Automated Storage and Retrieval in a Distribution Left. Master's Thesis, Massachusetts Institute of Technology, Cambridge, MA. May 2020.

Weiss, S., Schwarz, J.A. and Stolletz, R. 2019. The buffer allocation problem in production lines: Formulations, solution methods, and instances. IISE Transactions, 51(5), pp. 456–485. https://doi.org/10.1080/24725854.2018.1442031

Xi, S., Chen, Q., Smith, J.M., et al. 2020. A new method for solving buffer allocation problem in large unbalanced production lines. International Journal of Production Research, 58(22), pp. 6846–6867. https://doi.org/10.1080/00207543.2019.1685709.

Yuvamitra, K., Lee, J. and Dong, K. 2017. Value stream mapping of rope manufacturing: A case study. International Journal of Manufacturing Engineering, 2017, Article ID 8674187, 11 pages.

Appendix A

Manufacturing KPIs

The review and applications of key performance indicators (KPIs) are a core aspect discussed in this text. There are numerous manufacturing KPIs that reflect various practices in different industries. The list below provides these KPIs for the reader's reference, noting that variations exist in the descriptions, calculations, and units of KPIs.

All these KPIs are relevant to assessing the throughput performance of manufacturing operations to varying degrees. While some KPIs can directly be applied to manufacturing systems and processes, others may require adaptation. For more information about throughput KPIs and their selection, please refer to Chapter 2.

---

1   Actual to planned scrap ratio: This metric compares the actual scrap units produced to the planned scrap units.

$$\frac{\text{Actual scrap units}}{\text{Planned scrap units}} \times 100$$

2   Allocation efficiency: This metric measures the actual usage (in time) of a machine relative to its scheduled operating time.

$$\frac{\text{Actual busy time}}{\text{Planned operating time}} \times 100$$

3   Allocation ratio: This metric calculates the percentage of actual working time to the actual order execution time per unit.

$$\frac{\text{Actual unit busy time}}{\text{Actual order execution time}} \times 100$$

4   Availability (A): This metric represents the actual working/busy time that a machine or system is available to work in a given period (normally excluding setup time and planned downtime).

$$\frac{\text{Actual busy time}}{\text{Scheduled operating time}} \times 100$$

Manufacturing System Throughput Excellence: Analysis, Improvement, and Design, First Edition. Herman Tang.
© 2024 John Wiley & Sons, Inc. Published 2024 by John Wiley & Sons, Inc.
Companion website: www.wiley.com/go/Tang/ManufacturingSystem

5   Average unit contribution margin: This metric evaluates a product's contribution to cover fixed costs after considering all unit variable costs.

$$\frac{\text{Total revenue} - \text{Total variable costs}}{\text{Total volume of produced units}} \times 100$$

6   Avoided costs: This metric quantifies the savings realized from preventive maintenance activities.

$$(\text{Estimated repair cost} + \text{Production losses}) - \text{Preventive maintenance cost}$$

7   Blockage ratio: This metric measures the percentage of WIP that cannot proceed downstream due to interruptions from downstream processes during total production time.

$$\frac{\text{Blocked time}}{\text{Scheduled production time}} \times 100$$

8   Capacity utilization: This metric indicates the percentage of total availability of a plant (system) that is utilized within a specific time period.

$$\frac{\text{Capacity used}}{\text{Total available production capacity}} \times 100$$

9   Changeover time: This is the time required to transition a production line from one product to another (different) product by preparing the equipment or system.

10   Comprehensive energy consumption (or energy cost per unit): This metric represents the percentage of all the energy consumed in production relative to the number of units produced in a specific period.

$$\frac{\text{All energy consumed in production}}{\text{Number of units produced}} \times 100$$

11   Corrective maintenance ratio: This metric measures the percentage of corrective maintenance tasks performed in relation to all corrective and preventative maintenance activities within a given period.

$$\frac{\text{Corrective maintenance time}}{\text{All maintenance time}} \times 100$$

12   Cost per unit (or unit cost): This metric indicates the amount of money spent on each unit produced, typically calculated as the average cost of a batch of units. Total costs include all direct and indirect costs. This metric can be adopted for nonproduction areas, such as maintenance.

$$\frac{\text{Total manufacturing costs}}{\text{Total units produced}} \times 100$$

13   Critical process or machine capability index (C_{pk}): This metric measures the ability of a manufacturing process or machine to produce a product within the upper and lower specification limits (USL and LSL). It considers that the mean (μ) of a quality attribute for products with standard deviation σ may not be centered between the USL and LSL.

$$\frac{Min\,(\text{USL} - \mu, \mu - \text{LSL})}{3 \times \sigma}$$

14   Customer return (reject) ratio: This metric measures the percentage of products returned by customers for various reasons. It is often necessary to separate the rate based on different reasons.

$$\frac{\text{Units returned}}{\text{Total number of units shipped}} \times 100$$

15   Defect ratio: This metric calculates the percentage of defective units over the total units produced.

$$\frac{\text{Defective units}}{\text{Total units produced}} \times 100$$

16   Downtime ratio: This metric represents the ratio of downtime over the total scheduled production time. The definition of downtime can vary, including a combination of scheduled downtime and unscheduled downtime, unscheduled downtime only, or failure-related unscheduled downtime only.

$$\frac{\text{Downtime}}{\text{Total scheduled operating time}} \times 100$$

17   Employee turnover: This metric measures the percentage of employee separations relative to the average number of employees in a given period.

$$\frac{2 \times \text{Number of separations}}{\text{Number of employees at start} + \text{Number of employees at end}} \times 100$$

18   Equipment load ratio: This metric indicates the percentage of units produced relative to the equipment production capacity in product units in a given period. The metric may be based on time.

$$\frac{\text{Units produced}}{\text{Equipment production capacity (units)}} \times 100$$

19   Fall-off ratio: This metric measures the rate of fall-off units for a production operation in relation to the units produced initially.

$$\frac{\text{Fall off units}}{\text{Produced units}} \times 100$$

20   Finished good rate: This metric represents the percentage of good quality units produced relative to the consumed materials.

$$\frac{\text{Good quality units produced}}{\text{Consumed materials}} \times 100$$

21   First time yield (FTY): This metric measures the percentage of good quality product units (fully meeting the requirements) in the first process run only without reworks or repairs.

$$\frac{\text{Good quality units produced}}{\text{All units produced}} \times 100$$

22   Health and safety incidence rate (also known as the total case incident rate – TCIR): This metric calculates the number of work-related injuries per 100 full-time workers during a 12-month time period.

$$\frac{\text{Number of OSHA} - \text{Recorded injuries and illnesses} \times 200{,}000}{\text{Total employee hours worked}} \times 100$$

OSHA: Occupational Safety and Health Administration (US government agency)

Note: The constant value of 200,000 represents 100 employees working 40 hours per week, 50 weeks per year.

23   Inventory turns: This metric is a ratio of the output over the average inventory in a given period, based on physical unit counts.

$$\frac{\text{Cost of units sold}}{\text{Average inventory (units)}} \times 100$$

24   Lead time: This metric measures the total time it takes for customers to receive orders after they are placed.

$$\text{order process time} + \text{production lead time} + \text{delivery lead time}$$

25   Machine effectiveness: This metric represents the percentage of the product of the design cycle time and the units produced over the actual production time for a machine or system.

$$\frac{\text{Design cycle time} \times \text{Quality rate of units produced}}{\text{Actual production time}} \times 100$$

26   Machine utilization efficiency: This metric calculates the percentage of the difference between the actual cycle time (processing time) and unit setup time over the actual unit busy (process + delay) time for a machine or process per unit.

$$\frac{\text{Actual processing time} - \text{Setup time}}{\text{Actual busy time} \times \text{Number of units produced}} \times 100$$

27   Maintenance schedule compliance (MSC): This metric measures the percentage of scheduled maintenance work orders completed within a given timeframe.

$$\frac{\text{Number of completed maintenance orders}}{\text{Number of scheduled maintenance orders}} \times 100$$

28   Maintenance unit cost: This metric evaluates maintenance efficiency (sometimes excluding materials) within a specific time period.

$$\frac{\text{Total maintenance cost}}{\text{Number of units produced}}$$

29   Manufacturing cost as a percentage of revenue: This metric assesses the company's spending on manufacturing relative to total revenue.

$$\frac{\text{Total manufacturing cost}}{\text{Total revenue}} \times 100$$

30 Material yield variance: There are two common calculations. The first calculates the difference between the actual unit usage and the standard unit usage, multiplied by the standard cost per unit. The second calculation compares the actual material used to the expected material use.

$$(\text{Actual unit usage} - \text{Standard unit usage}) \times \text{Standard cost per unit}$$

$$\frac{\text{Actual material use}}{\text{Expected material use}} \times 100$$

31 Mean time between failures (MTBF): This metric represents the average time between two consecutive failures of a machine or system over a long time period, often measured in hours.

$$\frac{\text{Number of operating hours}}{\text{Number of failure events}}$$

32 Mean time to failure (MTTF): This metric measures the average time between two consecutive failures of a machine or system over a long time period, often measured in minutes.

33 Mean time to repair (MTTR): This metric calculates the average downtime (often repair time) for the failures of a machine over a long period, often measured in minutes.

$$\frac{\text{Total downtime}}{\text{Number of failures}}$$

34 Net equipment effectiveness (NEE): This metric is similar to OEE but replaces availability with the uptime ratio per unit $\left(\dfrac{\text{Actual processing time}}{\text{Scheduled production time}} \right)$.

$$\text{NEE} = \frac{\text{Actual processing time}}{\text{Scheduled production time}} \times \text{Performance} \times \text{Quality rate}$$

35 Noncompliance rate: This metric represents the number of products that do not comply with the regulatory rules of a region or country in a given period, such as a year.

$$\frac{\text{Noncompliance units or events}}{\text{Given period}}$$

36 On-time delivery: This metric calculates the percentage of products delivered on time to customers compared to the total volume of delivered products.

$$\frac{\text{Units on} - \text{Time delivered}}{\text{Total delivered units}} \times 100$$

37 Other loss: This metric represents the units lost not related to production, storage, and transportation, and may be expressed as an absolute value or as a ratio to the total units produced.

38 Overall equipment effectiveness (OEE) index: This metric is the product of availability, performance, and quality rate, serving as an integrated indicator (performance is sometimes referred to as effectiveness).

$$\text{Availability} \times \text{Performance} \times \text{Quality rate}$$

39　Overall labor effectiveness (OLE) index: This metric assesses human labor in a manner similar to OEE used for equipment, sharing the same equation but on the workforce.

$$\text{Labor utilization} \times \text{Performance} \times \text{Quality rate}$$

40　Overall throughput effectiveness (OTE) index: This metric compares actual productivity to maximum attainable productivity as a ratio.

$$\frac{\text{Actual throughput (units) from factory in total time}}{\text{Theoretical throughput (units)from factory in total time}}$$

41　Overtime ratio: This metric provides a relative measurement of overtime work in a period to show inefficiencies in scheduling, output, and/or staffing. Two common calculations are:

$$\frac{\text{Overtime work hours}}{\text{Regular work hours}} \times 100$$

$$\frac{\text{Overtime work hours}}{\text{Total work hours}} \times 100$$

42　Planned maintenance percentage (PMP): It is the percentage of planned maintenance tasks of all maintenance tasks in a time period.

$$\frac{\text{Planned maintenance hours}}{\text{Total maintenance hours}} \times 100$$

43　Process or machine capability index (C_p): This metric measures the ability of a manufacturing process or machine to produce a product within the upper and lower specification limits (USL and LSL). It assumes that the mean of a quality attribute for products with standard deviation σ is centered between the USL and LSL.

$$\frac{\text{USL} - \text{LSL}}{6 \times \sigma}$$

44　Production attainment: This metric measures the level of production achieved over a specific time period, e.g., per a week.

$$\frac{\text{Number of periods production target met}}{\text{Total time}} \times 100$$

45　Production loss rate: This metric represents the relationship between the units lost and the consumed materials during a specific time period.

$$\frac{\text{Quality lost}}{\text{Consumed materials}} \times 100$$

46　Production process ratio: This metric is the percentage of actual production time compared to the total time of actual order execution.

$$\frac{\text{Actual production time}}{\text{Actual order execution time}} \times 100$$

47   Quality buy ratio: This metric indicates the percentage of good quality units, including reworks, among all units produced.

$$\frac{\text{Good quality units}}{\text{Produced units}} \times 100$$

48   Quality ratio (rate): It is the overall percentage of good quality units over the total number of units produced, which may include reworked units.

$$\frac{\text{Good quality units}}{\text{Produced units}} \times 100$$

49   Rate of new product introduction: This metric measures the frequency of a company introducing new products to the market within a specific period.

$$\frac{\text{Number of new products}}{\text{Given period}}$$

50   Rate of return (ROR): It is a financial measure that assesses the performance of a capital investment over time.

$$\frac{\text{Current value} - \text{Initial value}}{\text{Initial value}} \times 100$$

51   Rolled throughput yield (RTY): It is the cumulative probability that a process with multiple sequential steps will produce a defect-free unit. $(1 - p_i)$ is the FTY (first time yield) of step i.

$$\text{RTY} = (1 - p_1) \times (1 - p_2) \times \cdots \times (1 - p_n)$$

52   Reportable environmental incidents: It refers to incidents related to air and water pollution, recycling, or other environmental issues that are required to be reported to the Environmental Protection Agency (EPA) in the US within a specific period.

53   Return on assets (ROA): This metric is a financial metric that measures the profitability of a company's assets for a period (e.g., annually), considering total assets including fixed assets and working capital.

$$\frac{\text{Net income}}{\text{Average total assets}} \times 100$$

54   Rework ratio: It represents the percentage of reworked units meeting quality requirements compared to the total units produced.

$$\frac{\text{Reworked units}}{\text{Total units produced}} \times 100$$

55   Schedule attainment (SA): It is an indicator of production output that considers both actual and planned production units and scheduled production time.

$$\frac{\dfrac{\text{Actual produced units}}{\text{Actual production time}}}{\dfrac{\text{Planned to produce units}}{\text{Scheduled production time}}} \times 100$$

56   Scrap ratio: This metric measures the rate of scrap units, which are not repairable, compared to the total number of units produced. This ratio can also be used for incoming materials and measured in terms of weight or monetary value.

$$\frac{\text{Scrap units}}{\text{Total units produced}} \times 100$$

57   Setup ratio (rate): This metric represents the relative loss of the value-adding opportunity for a machine or system due to setup time. There are two common calculations:

$$\frac{\text{Actual unit setup time}}{\text{Actual unit processing time}} \times 100$$

$$\frac{\text{Available time} - \text{Processing time}}{\text{Number of units produced}} \times 100$$

58   Setup time: It is the time required to prepare equipment or a system for its next production run after completing a run.

59   Starvation ratio: This metric is the percentage of WIP that cannot proceed from upstream due to interruptions in the upstream process, relative to the total production time.

$$\frac{\text{Starved time}}{\text{Scheduled production time}} \times 100$$

60   Storage and transportation loss rate: It represents the relationship between the lost quantity during storage and transportation and the consumed materials (or total incoming materials).

$$\frac{\text{Loss during storage and transportation}}{\text{Consumed materials}} \times 100$$

61   Supplier quality rate: This metric is the quality ratio of raw materials and parts received from a supplier or all suppliers over a specified period.

$$\frac{\text{Good quality materials received}}{\text{All incoming materials received}} \times 100$$

62   Technical efficiency: This metric represents the percentage of the actual production time over the sum of the actual production time and the malfunction-caused interruptions and delays due to technical reasons.

$$\frac{\text{Actual production time}}{\text{Actual production time} + \text{Down time}} \times 100$$

63   Throughput rate (TR) or simply throughput: This metric represents the ratio of good quality units produced over the execution time. This includes good quality units after rework and repair. The production time may be the order execution time in certain scenarios.

$$\frac{\text{Good quality units}}{\text{Actual production time}}$$

64   Throughput time: It is the total time needed for a system to produce a unit.

$$\text{Process end time} - \text{Process start time}$$

65   Total effective equipment performance (TEEP): It is a productivity metric that considers system availability (A), operating speed performance (P), quality rate (Q), and time utilization (U).

$$\text{TEEP} = A \times P \times Q \times U = \text{OEE} \times U$$

66   Total manufacturing cost per unit excluding materials: It is a metric that calculates the fixed costs associated with operating a factory or system per unit produced, excluding the cost of materials.

$$\frac{\text{Total manufacturing cost}}{\text{Units produced}}$$

67   Unscheduled downtime ratio: This metric represents the percentage of unscheduled downtime relative to scheduled downtime in a given period.

$$\frac{\text{Unscheduled downtime}}{\text{Scheduled downtime}} \times 100$$

68   Volume attainment (VA): This metric is a direct measurement of the quantity of production output achieved in a given period.

$$\frac{\text{Actual produced units}}{\text{Planned to produce units}} \times 100$$

69   Worker efficiency: This metric represents the percentage of a worker's actual work time on production orders compared to their scheduled work time (or actual attendance time).

$$\frac{\text{Actual work time}}{\text{Scheduled work time}} \times 100$$

70   Work in process (WIP): It refers to the number of partially completed product units within a system, either on average or at a specific moment.

Appendix B

Acronyms, Abbreviations, and Notations

This text uses numerous acronyms and abbreviations. They are fully spelled out upon their initial introduction but may not be spelled out in every subsequent chapter. Readers can refer to this Appendix, as necessary. It is important to note that there may be variations in spelling across different sources.

5D	five disciplines (problem-solving)
5S	sort, set in order, shine, standardize, and sustain
8D	eight disciplines (problem-solving)
A	operational availability
A3	A3 template (problem-solving)
AGV	automated-guided vehicle
AI	artificial intelligence
AIAG	Automotive Industry Action Group
ALB	assembly line balancing
ALM	asset lifecycle management
AM	autonomous maintenance
Andon	system to notify other personnel of quality or process problems
AR	auto regressive
ARMA	autoregressive moving average
ARP	aerospace recommended practice
AS	aerospace standard
A_{sa}	standalone availability
ASQ	American Society for Quality
ASRS	automated storage and retrieval system
ATO	assemble-to-order
ATS	assemble-to-stock
B_1	buffer impact factor
B_2	buffer impact factor
BAP	buffer allocation problem
C/C	cross car (a coordinate direction)
$C_{equipment}$	equipment cost

Manufacturing System Throughput Excellence: Analysis, Improvement, and Design, First Edition. Herman Tang.
© 2024 John Wiley & Sons, Inc. Published 2024 by John Wiley & Sons, Inc.
Companion website: www.wiley.com/go/Tang/ManufacturingSystem

$C_{facility}$	facility cost
CI	continuous improvement
C_j	consequence rating of failure j
C_{labor}	labor costs
$C_{material}$	cost of materials
CONWIP	constant work in process
COQ	cost of quality
$C_{overhead}$	overhead cost
C_p	process capability index
C_{pk}	critical process capability index
CT	cycle time
CT_a	actual cycle time
CT_d	design cycle time
C_{total}	total cost
CT_u	cycle time loss due to upstream operations
$C_{utility}$	utility costs
CV	coefficient of variation
D	failure detection rating (in FMEA)
DES	discrete event simulation
DFQ	design for quality
D_j	average duration of failure j
DMAIC	define, measure, analyze, improve, and control
DOD	US Department of Defense
DOE	US Department of Energy
DOI	digital object identifier
DRBFM	design review based on failure modes
ED	emergency department (in hospitals)
f	frequency of downtimes
F	future value
F/A	fore/aft (a coordinate direction)
FEA	finite element analysis
FIFO	first in first out
F_j	frequency of failure j
FMEA	failure mode and effect analysis
FPY	first-pass yield
FTA	fault tree analysis
FTQ	first-time quality
GD^3	good design, good discussion, good dissection
GM	General Motors
i	interest rate
IATF	International Automotive Task Force

IEC	International Electrotechnical Commission
IIHS	Insurance Institute for Highway Safety
I_j	reliability importance of j
ISO	International Organization for Standardization
JIT	just in time
JPH	job per hour
KPI	key performance indicator
LOC	logistic operating curve
LPA	layered process audit
LSL	lower specification limit
LSS	lean six sigma
MA	moving average
MARR	minimum acceptable rate of return
Max	maximum
MIL-HDBK	military handbook
Min	minimum
min	minute
MSC	maintenance schedule compliance
MTBF	mean time between failures
MTM	methods-time measurement
MTTF	mean time to failure
MTTR	mean time to repair
n.d.	no date
NAN	non-value-added but necessary
NASA	National Aeronautics and Space Administration
NEE	net overall equipment effectiveness
NHTSA	National Highway Traffic Safety Administration
NUREG	Nuclear Regulatory Commission
O	failure occurrence rating (in FMEA)
OEE	overall equipment effectiveness (efficiency)
OEE_{sa}	standalone OEE
$OEE_{sa\text{-}simplified}$	simplified standalone OEE
OEE_w	weighted OEE
OLE	overall labor effectiveness
OPE	overall production effectiveness
OTE	overall throughput effectiveness
P	performance (in OEE)
P	present value
P	probability
p	quality issue rate
p_a	actual produced units

PAF	prevention-appraisal-failure
PDCA	plan, do, study, and act (problem-solving)
PDPC	process decision program chart
PI	priority identification number
PMP	planned maintenance percentage
PO	production order
PO	purchase order
PS	problem-solving
P_{sa}	standalone performance (in OEE)
p_u	speed loss due to upstream operations
P_{WIP}	probability of having WIP units in a buffer
Q	quality (in OEE)
QC	quality control
QFD	quality function deployment
QMS	quality management system
Q_{sa}	standalone quality (in OEE)
R	range
R	reliability
R_A	rating of availability (in OEE)
RCM	reliability-centered maintenance
R_j	component's reliability
ROI	return on investment
R_p	rating of performance (in OEE)
RPN	risk priority number (in FMEA)
R_Q	rating of quality (in OEE)
R_{sys}	system's reliability
RTY	rolling throughput yield
RWIP	WIP change rate
S	failure severity rating (in FMEA)
s	second
SA	schedule attainment
SAE	Society of Automotive Engineers
SIPOC	suppliers, inputs, process, outputs, and customers
S_j	severity of failure j
SMART	specific, measurable, achievable, relevant, and time-bound
S_p	system performance
SWOT	strengths, weaknesses, opportunities, and threats
t	time
T	work time
t_a	actual operating (up) time
t_b	blocked time

TEEP	total effective equipment performance
t_{load}	time needed to load a part
TOC	Theory of Constraints
t_p	planned production time
TPM	total productive maintenance
TPY	throughput yield
TQM	total quality management
TR	throughput rate
TR_{gross}	gross throughput rate
TRIZ	theory of inventive problem-solving
TR_{net}	net throughput rate
TR_{sa}	standalone throughput rate
t_s	starved time
TT	throughput time
t_{travel}	part travel time on a conveyor
t_u	actual production (up) time
t_{unload}	time needed to unload a part
U	time utilization (in TEEP)
U/D	up/down (a coordinate direction)
u_a	actual units produced
u_g	good units produced
u_p	planned units in uptime
USL	upper specification limit
u_u	defective units due to upstream operations
VA	value added
VA	volume attainment
VDA	German Association of the Automotive Industry
VDI	Association of German Engineers
W	waste
w_A	weight for A in OEE
WIP	work in process
w_P	weight for P in OEE
w_Q	weight for Q in OEE
ΔA	availability change (in OEE)
ΔP	performance change (in OEE)
ΔQ	quality change (in OEE)
Δt	time interval
ΔWIP	WIP change
$\Delta \mu$	change of mean
$\Delta \sigma$	change of standard deviation
λ	failure rate
μ	arithmetic average or mean
σ	standard deviation

Appendix C

Answers to Exercises

There is a keyword (in parentheses) associated with each problem as a reference. Solution worksheets in MS Excel are available to instructors.

Chapter 1: Throughput Concept

1.1   (availability) 90.7%; 94.0%

1.2   (productivity) 15 units/employee; 0.358 units/hour

1.3   (cost) 31.7%

1.4   (WIP) 225 units

1.5   (WIP) 7.7 units

1.6   (throughput rate) 75.0 JPH

1.7   (standalone) 95.8%; 74.57 JPH

Chapter 2: System Performance Metrics

2.1   (availability) 100.2%; 94.33%

2.2   (performance) 98.67%

2.3   (throughput loss) 8.8 units/shift

2.4   (OEE) 83.33%

2.5   (TEEP) 56.7%

2.6   (OEE estimation) 0.8%

2.7   (weighted OEE) 93.58%

2.8   (standalone OEE) 92.8%

2.9   (throughput loss) $52,500

Chapter 3: Bottleneck Identification and Buffer Analysis

3.1   (bottleneck) workstation 3 – 62.07 JPH

3.2   (active period) subsystem 1 – 95%

3.3   (turning point) operation 3 – 70 minutes of starved and blocked time

3.4   (buffer min) 2.1 units and roundup to 3 units

3.5   (bottleneck) subsystem 2

3.6   (buffer effect) 10 minutes for upstream; 5 minutes for downstream

3.7   (WIP rate) alarm 37.5 units/minutes; 40 units/minutes

Manufacturing System Throughput Excellence: Analysis, Improvement, and Design, First Edition. Herman Tang.
© 2024 John Wiley & Sons, Inc. Published 2024 by John Wiley & Sons, Inc.
Companion website: www.wiley.com/go/Tang/ManufacturingSystem

Chapter 4: Quality Management and Throughput

4.1   (break-even) 26.67 weeks

4.2   (MARR) 16.4%

4.3   (MARR) $3233.55

4.4   (RTY) 0.808

4.5   (parallel mean) 14.47

4.6   (parallel variation) 6.46

Chapter 5: Maintenance Management and Throughput

5.1   (failure rate) 99.84%; 99.92%

5.2   (priority) failure mode C

5.3   (MTBF) 9.78 hours

5.4   (MTTR) 11.41 minutes

5.5   (failure rate MTBF) 6.45%

5.6   (availability) 98.41%

5.7   (PMP) 79.2%

5.8   (total costs) $335; $330

5.9   (priority OEE) 1 – 24 and 2 – 20

Chapter 6: Throughput Enhancement Methodology

6.1   (variation) 4.82%

6.2   (curve fitting) $y = 45.042e^{-0.552x}$ (using MS Excel)

6.3   (cycle time – serial) 58 seconds

6.4   (cycle time – parallel) 28.9 seconds

6.5   (evaluation) B – 36.75

6.6   (gap analysis) tasks 6 – 0.9, 1 – 0.8, and 4 – 0.6

Chapter 7: Analysis and Design for Operational Availability

7.1   (reliability – serial) 80.7%

7.2   (reliability – parallel) 81.7%; 17.4%

7.3   (reliability – hybrid) 66.66%; 21.93%

7.4   (bottleneck) station 5 – 0.868 importance

7.5   (slow backup) 77.0 JPH; 75.2 JPH

7.6   (auto backup) 97.4%

Chapter 8: System Design for Throughput Assurance

8.1   (no buffer) 50.40 JPH

8.2   (buffer) 52.47 JPH

8.3   (workload) 24.0%

8.4   (design) no – 59.8 JPH, 3.56 JPH

8.5   (reliability design) 42.3, 44.1, 43.7, 42.3, 43.2, 42.8, and 44.6 seconds

8.6   (OEE design) 55.8, 52.8, 54.0, 53.4, 55.2, and 54.6 seconds

Epilogue

This book provides a comprehensive guide to advancing manufacturing excellence, covering a wide range of topics such as data-driven approaches, bottleneck-focused processes, financial analysis, integration of quality and maintenance, and proactive design. Familiar methods reinforce your knowledge, while other approaches encourage you to think and implement throughput solutions in new and different ways.

This book presents adaptable frameworks rather than one-size-fits-all recipes – creative applications tailored to each production system's unique needs. By implementing the insights and frameworks presented here, you can elevate your mindset and actively pursue manufacturing excellence to drive impactful changes.

Thank you for joining me on this journey to enhance manufacturing throughput. I hope this book empowers you to achieve even greater success and sparks creative thinking about solutions. Your feedback and stories are invaluable; I would love to hear from you.

Manufacturing System Throughput Excellence: Analysis, Improvement, and Design, First Edition. Herman Tang.
© 2024 John Wiley & Sons, Inc. Published 2024 by John Wiley & Sons, Inc.
Companion website: www.wiley.com/go/Tang/ManufacturingSystem

Index

Manufacturing System Throughput Excellence: Analysis, Improvement, and Design, First Edition. Herman Tang.
© 2024 John Wiley & Sons, Inc. Published 2024 by John Wiley & Sons, Inc.
Companion website: www.wiley.com/go/Tang/ManufacturingSystem

Printed and bound by CPI Group (UK) Ltd, Croydon, CR0 4YY

16/04/2025

14658353-0002